AUDIO SAMPLING

A Practical Guide

Sam McGuire

Roy Pritts

ELSEVIER

AMSTERDAM • BOSTON • HEIDELBERG • LONDON
NEW YORK • OXFORD • PARIS • SAN DIEGO
SAN FRANCISCO • SINGAPORE • SYDNEY • TOKYO

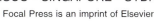

Focal Press is an imprint of Elsevier

Focal Press

Acquisitions Editor: Catharine Steers
Publishing Services Manager: George Morrison
Project Manager: Mónica González de Mendoza
Assistant Editor: David Bowers
Marketing Manager: Marcel Koppes
Cover Design: Eric Decicco

Focal Press is an imprint of Elsevier
30 Corporate Drive, Suite 400, Burlington, MA 01803, USA
Linacre House, Jordan Hill, Oxford OX2 8DP, UK

 Recognizing the importance of preserving what has been written, Elsevier prints its
books on acid-free paper whenever possible.

Library of Congress Cataloging-in-Publication Data
McGuire, Sam.
    Audio sampling : a practical guide / Sam McGuire.
        p. cm.
    Includes bibliographical references and index.
    ISBN 978-0-240-52073-5 (pbk. : alk. paper)   1. Software samplers.   I. Title.
    ML74.3.M43 2008
    784.13'4—dc22

                                                                        2007025666

British Library Cataloguing-in-Publication Data
A catalogue record for this book is available from the British Library.

ISBN: 978-0-240-52073-5

For information on all Focal Press publications
visit our website at www.books.elsevier.com

07 08 09 10 11 12 13   10 9 8 7 6 5 4 3 2 1

Printed in the United States of America

Working together to grow
libraries in developing countries

www.elsevier.com | www.bookaid.org | www.sabre.org

ELSEVIER   BOOK AID
            International   Sabre Foundation

To the memory of Roy Pritts (1936–2007).

ACKNOWLEDGMENTS

Thank you to Lorne Bregitzer (MIDI master) and Storm Gloor (industry guru) for your great additions. Thank you to Martin Russ for showing no mercy. It really made a difference. Thank you to Sarah Smith for the super sweet illustrations. Thank you to Catharine Steers, David Bowers, and everyone else at Focal for all the hard work (and patience).

—Sam McGuire

FOREWORD

The day's e-mails contain a name I haven't seen for awhile. With a smile I recall Sam McGuire, a top student in the recording program here at BYU just a few short years ago. While most graduates can't wait to enter the world of work—usually finding a staff engineer position or opening a modest project studio—Sam surprised all of us by skipping all that and heading straight to Colorado for graduate work in audio engineering at the University of Colorado Denver.

Not only was Denver a good academic fit for someone coming from BYU (both programs reside in the School of Music and require a solid musical grounding as well as rigorous work in the related technology), but this choice would also put Sam in direct contact with Roy Pritts.

As the new millennium dawned, Roy Pritts was arguably the most visible national personality promoting audio education at the university level. In addition to heading the Denver master's program, he had a national platform within in the influential Audio Engineering Society (AES), heading its education arm. In that capacity I had first met him when he visited BYU's Sound Recording Technology program and gave a well-attended Saturday afternoon master class. Months later we would say hello again at AES in San Francisco. After that there would be occasional phone calls and e-mails, comparing programs or recommending students, or a chance hello at some airport.

Sam McGuire flourished in Denver, as he had at BYU, and snagged a good position recording and teaching audio at Appalachian State University after finishing his degree. The next thing I heard, Sam had applied and been selected for the position Roy Pritts was vacating upon retirement from UCD, now known by the ponderous acronym UCDHSC, for

University of Colorado Denver and Health Sciences Center (a name too long for a song title and hard to get used to, like Charlie's Barbershop and Oriental Food Store, an improbable venture that once graced State Street in Salt Lake City). So, in less than a decade after earning his undergraduate diploma, Sam McGuire had not only received his master's degree in Recording Arts, but also reached a sort of pinnacle in audio education, following in the footsteps of his much-admired mentor, the founder and head of the Denver Recording Arts program, Roy Pritts.

Sam's e-mail announces he is writing a book for Focal Press on sampling. (Way to go, Sam! I thought.) There's more: Roy Pritts will be contributing some key material (great again), and tucked in at the end: would I contribute the foreword?

I said I was honored and, yes, I'd be happy to.

Sampling—it seems like the blink of an eye since the Beatles' record "Strawberry Fields Forever" introduced the sound of analog tape samples played on the Mellotron. Digital was not far behind, with the Fairlight and other pioneering samplers. Soon after, it seemed everyone was bringing a sampler of choice along with their favorite synths to the recording studio.

And then, not long ago, music production guys were heard saying, "I just unhooked my last outboard sampler," and the era of computer-based onboard sampling seemed to take over. Not one but several prepackaged orchestral sample libraries had found substantial pockets of buyers, and gradually even the sounds we all thought couldn't be effectively sampled were successfully introduced into the general market: strummed and picked acoustic guitars, handmade ethnic flutes. Suddenly sample libraries seemed almost like an embarrassment of riches.

My sampling reverie is interrupted by the phone. It's producer and folk music impresario Clive Romney. "I'm doing a sampling project for Gary Garritan up in Seattle [Garritan Personal Orchestra, etc.]," Romney says, and tells me he's already captured an amazing list of folk percussion instruments from exotic places: instruments and places I've mostly never heard of. "So," Clive focuses in on his errand, "would you know anyone who has a sarangi, or any wasembe or kayamba rattles?" I'm blown away that someone I know so well is right in the thick of sample library production. "Sorry, can't help with any of those," I manage, "but if you get into exotic stringed instruments, we have a Moroccan lute and a Finnish kantele you could use."

I change the subject: "Hey, before you hang up, how many production folks are still using outboard samplers?" (I'm thinking ahead to Sam's foreword.) "Well, me, for starters," he says, and names his favorite Roland box.

Sampling, then, is a big world, with outboard samplers, computer plug-ins, prepackaged third-party library samples, and the creation of custom user-created sounds—it goes on and on.

Sam McGuire and co-author Roy Pritts are the experts. They're thoroughly armed with the theory and technology of sampling, taking whatever it is you already know and filling in the details around it. They are my favorite kind of music-technology gurus: they know the present, appreciate the past, and have pretty accurate ideas about the future.

Get ready for a good ride.

Just after completion of the book manuscript, and somewhere in the middle of preproduction, word came that Roy Pritts had suddenly passed away. This book may be the last writing project he worked on. Thanks, Roy, not just for this book, but for the influence you've had on so many students such as Sam McGuire, and for your years of volunteer work at AES. Guys like you are so hard to find—and such a tough act to follow.

Ron Simpson
General manager, Tantara Records
Division coordinator, Media Music, Brigham Young
University School of Music

PREFACE

This project started as a quest to find a book on sampling. When I couldn't find any comprehensive texts on the creation of sampled instruments, I began to formulate the outline for this book. One of the most exciting parts about this project has been developing a presentation that takes advantage of really great technology. Besides the text and illustrations included here, we've prepared a thorough web resource. It includes video tutorials, audio examples, and current product descriptions/comparisons. We debated about whether to use a CD-ROM, but in the end the website seemed like a better long-term offering. The goal of the website is to give you the most value possible. Instead of including media content that would be outdated in a matter of months, we are offering the content in a continuously updatable format. In the text you will find graphic indicators that point toward content on the website. These are designed to help you move back and forth between the text and the website.

In addition to a basic explanation of sampling, there are sections on several related topics. From a section on how to connect different equipment, to good ear training practices, the goal has been to put the proper tools in your hands to create excellent sampled instruments and to use them well. If there is only one thing that I hope you get from this book, it is an understanding that simply knowing how to do something never replaces true skill. Knowledge will never replace wisdom. As you embark into the world of sampling, strive for excellence in sound quality and instrument implementation. I hope you have as much fun reading this as we had in writing it!

—Sam McGuire

CONTENTS

INTRODUCTION TO SAMPLING 1

1.1 What is sampling?

This book is an introduction to the creation of virtual instruments through sampling. Sampling is the process of recording a sound source one part at a time, each part of which is then imported into a sampler. Typical parts include each note recorded from a musical instrument or singer. However, a sampler can be used with any recorded audio files. You could record yourself talking and each word could be a part of the sampled instrument. The sampler is used to organize the files into a playable instrument. A sampled instrument can be played using any MIDI controller (Figure 1.1), such as a MIDI keyboard.

Sampled instruments are used in many different phases of the music production process. From the initial song creation to the final record production, samplers have made a significant impact on how musical tasks are accomplished. For example, samplers are often used during the demo creation process. A sampler can be used to play all of the instrumental parts of a demo. Using a sampler, instead of hiring musicians, can be a huge money saver because you can play all of the parts without having to pay other people. Using a sampler can also unleash the creative process. Even after a demo is created using a sampler, it can be easily tweaked and altered. There is very little turn-around time from hearing the mixed demo to making alterations. If you use actual musicians in the demo process, then they have to be called in for further recording sessions if changes need to be made. Using a sampler means freedom from relying on other people and

Figure 1.1 MIDI
Controller.

their schedules. Throughout this book, we explore the many different parts of sampling.

Samplers are used to replicate sound sources with the potential for highly realistic results. Sampling is accomplished through several phases.

- The sound source is recorded one part at a time. An example of this might include each note produced by a pitched instrument or a collection of percussive drum hits.
- The recordings are edited.
- The resulting files (samples) are mapped into the sampler. The mapping process involves assigning each sample to a zone that is associated with either a single key or a range of keys on a keyboard.
- Finally, a MIDI sequencer or controller is used to trigger the samples.

While this is a simplified list, it accurately describes the core sampling process. Modern samplers have a wide range of features that are described later in this chapter. Some of the features are simple, while others are quite complex. This book is designed to explain each of these features in enough detail so that you can create your own virtual instruments. This chapter is also useful if you are trying to decide which

As we move through this book, look for website information icons. These icons indicate that there are additional resources, including explanations, illustrations, and/or associated audio/video clips on www. practicalsampling.com

type of sampler to purchase and use. Even if you are only interested in using sample libraries and not interested in creating your own sampled instrument, this book will help you to understand the sampling process and to use your samples in an effective manner.

Sampling is an effective tool because it uses a representative portion of a sound source to re-create the sound in an easily storable and recallable format. While a sampler is not able to reproduce most sound sources with the same level of variation as the original, it is able to achieve realistic results. The level of detail depends on how many samples are recorded and mapped into the sampler. It also depends on the specific sampler capabilities. Some samplers have features that allow for a higher level of realism. A realistic performance using a sampler depends less on the samples and the sampler capabilities and more on the performer's abilities. In the end, the sampler is still only an instrument that can make no music by itself. The realism and musicality of a sampler depends on the performer.

Another related technology that is sometimes confused with sampling is *modeling*. Just like sampling, modeling is an extremely realistic method of creating virtual instruments. The creation of a modeled instrument is accomplished by taking a large number of detailed measurements of a sound source. This data is then programmed into a software application that can be used in the same way as a sampler. The process of creating a modeled instrument is similar in concept to sampling, but there are no recordings made or samples used. Instead, the instrument is created through mathematics and computer programming.

While modeling has come a long way, in many cases instruments created through sampling rather than modeling give better results. Even though a sampler might have less potential for natural variation than a modeled instrument, the sampled instrument's strength lies in using actual recordings of the original sound source. Samplers are more accessible to end users who can tweak and create their own virtual instruments. The sampling process is not limited to the accurate re-creation of sound sources but can also be used to form new sounds and instruments by combining and altering recorded samples. Sampling is a flexible and highly creative tool for everyone wanting to create their very own instrument.

MIDI (musical instrument digital interface) is a critical concept to understand when discussing samplers. While this text makes no attempt to fully explain MIDI (but see Section 2.13 in Chapter 2 and Sections 7.1 and 7.7 in Chapter 7, with additional resources listed), it is a very important part of sampler control. MIDI is a data exchange protocol used to send control information to samplers. Without MIDI, a sampler is only a fairly fancy audio file storage system. MIDI is used to trigger the audio files and create a performance. For clarity and simplicity, this text uses the terms *triggered* and *played* as synonyms when referring to a sampler being controlled through MIDI. For example, saying a sample was triggered is the same thing as saying the sample was played. Both refer to MIDI data controlling what the sampler does. In addition, while there is more than one type of MIDI controller, this text uses a piano-style MIDI keyboard as the default type, unless specified otherwise.

1.2 A sampler survey

There are a number of sampling technologies available, ranging from simple sample playback tools to complex sample creation machines. This large range of samplers, while providing options for a variety of skill levels and interests, creates confusion about which one should be used in different situations. This section covers the primary sampler categories and provides insight into practical applications for each category.

1.2.1 Sample players

While nearly all samplers (Figure 1.2) have the ability to play prerecorded samples, there are some that are limited to playback and have no sample creation options. Many sample

Figure 1.2 Example of Sample Player.

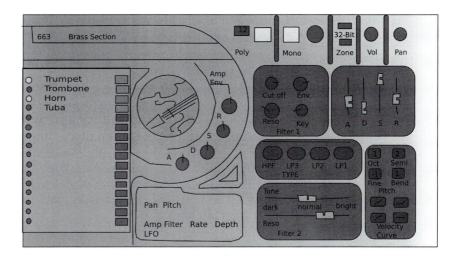

players are extremely powerful with high-quality sounds; they just can't be tweaked past a certain point. For example, you may not be able to edit individual sample files. Hardware sample players commonly are keyboards that have a set of standard sounds, but you can also find rack-mountable units that provide sample libraries but have no music keyboard built in. Software sample players typically have a primary focus, such as orchestral instruments, choirs, or traditional rock instruments. Some sample players are focused on one specific group of instruments, whereas others have a more general mix of different instruments. Sample players also come at a variety of prices. You can find software sample players with limited libraries that are very inexpensive, but sample players that come with large sample libraries can cost many times more. Not only do the sample players with large libraries cost a lot, but you might also need to have a dedicated hard drive to store the samples, adding further costs.

Website Link

Data storage comparison

You might be asking yourself why you would want to buy a sample player that has limited functionality in comparison to a full-blown sample recorder/editor/player. There are several valid reasons, the first being availability. Some sample libraries are not available except as a part of a sample player. Some manufacturers prefer to keep their samples and programming in a proprietary player format that is not open to "under the hood" tweaks and alterations. If you want to use these samples, you must purchase their specific sample player. Some of the very best sample libraries are available only in sample player format.

Another valid reason to use a sample player is if you have no need for a full-blown sampler with complex features that you will never use. If you don't plan to create your own instrument, then having a sampler might be too much. In this case, you can simply find a sample player with the sounds that you would like to use, and off you go. Be aware that several digital audio workstations (DAW) come with very capable sample players.

DAW sampler feature comparison

These particular players are seamlessly integrated into their specific host DAW. The reason they are included is to extend the software's functionality and give the end user a reason to purchase that particular software among many different choices.

If you are using samples in a live setting, a good keyboard with built-in samples might be a desirable option. Having the samples built into the keyboard reduces the need for a computer and audio interface and also reduces potential problems that might occur during live performances. Using a single performance keyboard on stage reduces the amount of equipment and cabling required. By reducing the amount of equipment, you likewise reduce potential problems caused by such things as cables failure and incorrect connections.

Some entry-level keyboards are limited to the built-in sounds, but other keyboards have expansion slots for importing additional instrumental samples. Some sample players offer the ability to add more sounds from third-party manufacturers, but other sample players do not. You need to research the accepted formats for your specific sample player because not

Sample format comparison

all sample libraries are of equal quality. You might be limited to certain library types.

1.2.2 Hardware samplers

While hardware samplers (Figure 1.3) are no longer the primary sampler format, they are still being developed and can be used to create complex sampled instruments. The most popular modern hardware samplers come as part of keyboard and drum-pad workstations. These workstations typically have at least a built-in controller (keyboard or drum pads), MIDI input/output (I/O), microphone/line level inputs, master outputs, and a software-based sampler interface. Having all of the necessary functions in a single box is very convenient. There are also a number of samplers that are rack-mountable boxes that have MIDI I/O, microphones/line level inputs, and master outputs.

There are several advantages and disadvantages to using hardware samplers. The primary advantage is that you have all the necessary tools (and often many more) for sampling. You don't have to worry about compatibility between different parts of the sampling process; everything has been tested and designed to work efficiently. Another advantage is for the performing musician. Like the sample players described earlier, they don't have a lot of separate components that

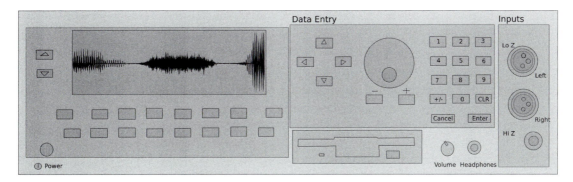

Figure 1.3 Example of Hardware Sampler.

7

have to be carried around and set up for every concert. Hardware sampler manufacturers take pride in offering reliable systems with high-quality sound libraries and a lot of sounds. Another advantage is the price. While some of these samplers are not cheap, they are still less expensive than an equivalent computer, audio interface, MIDI controller, and software package.

As for the disadvantages, you will find that the best part of a hardware sampler for some people is the worst part for someone else. While a hardware sampler might have all of the needed components to create a sampled instrument, it may not offer as much flexibility as a software sampler. The software that is included with the typical hardware sampler is usually not expandable with third-party effects. This means you are limited to the specific software of the hardware sampler. In addition to limited flexibility, the controls are often harder to use because there is less screen space to see parameters and editing screens. The features of hardware samplers, while undoubtedly advanced, are often buried among a lot of different page layers on a tiny screen.

Hardware samplers can often be expanded to increase their sounds and memory. This and other basic features are very similar to software samplers. The most obvious difference is in the packaging, but in the following section on software samplers, you will see other significant differences that demonstrate the increased power of the software sampler.

1.2.3 Software samplers

In many ways, software samplers (Figure 1.4) are the same as their hardware counterparts. Most of the differences are due to the fact that software samplers are operated on computers and they can take advantage of the various resources that come with computers. Software samplers are not inherently better or more powerful than hardware samples, and in some cases using a software sampler opens the door to many potential problems.

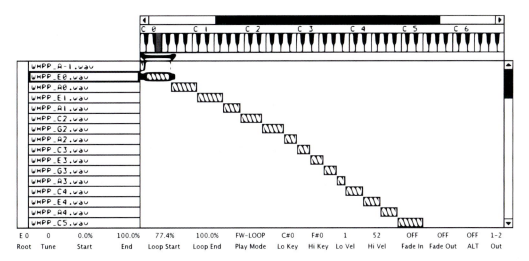

Figure 1.4 Example of Software Sampler.

One of the obvious advantages of using a software sampler is the screen space. The majority of hardware samplers have tiny screens that have poor resolution and complex menu structures. While there are hardware samplers that allow you to hook up an external display so you can better see what you are doing, software samplers always utilize your computer's display. Although computer screens come in a variety of sizes, nearly all of them are larger than the screens on hardware samplers. Having a larger screen workspace can make a huge difference when creating sampled instruments with hundreds of individual samples.

Comparison of digital audio editors

Another advantage to using a software sampler is the ease of collaboration with industry-standard digital audio editors.

While hardware samplers can be used to record and edit your source, they rarely have the functionality and expandability inherent in computer-based digital audio systems. The ability to record, edit, add effects from third-party manufacturers, and then drop the sample file into the sampler is a streamlined and easy process when using most software samplers.

While these are not the only advantages, they summarize the software sampler's universal advantages while avoiding the specific advantages of individual software samplers. There are samplers that have very advanced features in both the software and hardware categories, but there are limited examples of both as well. So, the question often becomes not which type of sampler to use, but which specific sampler to use.

To be fair, you should also be aware of some of the possible disadvantages of software samplers. One primary concern stems from the sharing of computer resources with multiple software programs. Not only does your computer run your sampler; it also runs your Internet browser, your word processor, your photo editor, and whatever else you want to install. When you use software from a multitude of companies, you run the risk of conflicts and unpredictable behavior. When you update your computer's operating system, it can make your sampler software obsolete. It is recommended that you add or change software very cautiously. If your budget allows, you should consider dedicating a single computer to music creation. This is standard practice for audio professionals. With a hardware sampler, you have a unit that is dedicated to one purpose and therefore is fundamentally more stable. While hooking a hardware sampler into an audio system is slightly more complex and requires additional cables, it might be worthwhile to keep your computer system free from the sampler process and the baggage that comes along with it.

Sampler list and comparison

As you can see, the choice you make when purchasing a sampler is not clear-cut. You have to weigh your options carefully and prioritize among the various functions. Sometimes the choice is easy because a sampler might come bundled with other software that you buy, and other times there is more than one perfectly good choice. As you read the later sections in this chapter dealing with feature expectations, you are presented with information to help make an educated decision regarding sampler choice.

Figure 1.5 Mellotron.

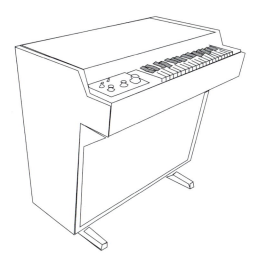

1.2.4 *Tape-based sampling*

While analog sampling is no longer a realistic option and there are no tape-based systems currently under development for large-scale commercial release, you should be aware that they were once the primary method for sampling. The Chamberlain and Mellotron (Figure 1.5) sample players used analog tape as the primary storage format for the sound sources. Each key of a keyboard was hooked to a piece of tape that would play back when the key was pressed. The process of recording sounds onto these tape segments and then maintaining consistency was not an easy process. Still, these samplers were used in a variety of settings and provided the only source of sampling for a number of years. See the historical chapter (Chapter 8) for more information on analog samplers.

1.2.5 *Miscellaneous utilities*

There are a number of utilities that add functionality to your sampler. A good example of this is a software utility that can create a fully sampled instrument by sending every possible trigger note to a keyboard or instrument that is controllable through MIDI. This type of utility records the output audio samples from the source and appropriately maps and layers the files into a sampled instrument. While an automatic system such as this is extremely useful, there is usually still a lot of tweaking to do to get the instrument just right. Utilities like this can be a huge time-saver in certain situations by automatically recording and splitting many different files across the keyboard. The purpose of this type of utility is not for you to hook it up to your friend's instruments and steal the sounds, but more for the musician who has a lot of hardware modules and wants to be able to use their sounds in a computer environment without having to carry a lot of equipment around.

Other utilities that can be a great help include file management software to help keep track of the many source files that you will be recording, file conversion software for bulk audio format conversions, and just about any other software that you can find that will help make the sampler creation process more efficient. Software such as Microsoft Word or Excel are good examples of tools that you can use to keep organized during the sample creation process.

Utilities

1.3   Basic sampler expectations

Now let's take a look at what you can and should expect from a modern sampler. We start with the basic and universal features that exist in all professional samplers.

1.3.1   Basic editing features

At the core of every sampler is a group of recorded audio files. These files may be recorded directly into the sampler, or imported into the sampler after being recorded into an external recorder. In either case, the individual audio files need to be prepared using a variety of editing techniques and tools.

1.3.2 Editing the audio file length

Most samplers allow you to change the length of the individual audio files. This involves adjusting the beginning and the ending of the audio files. Many samplers will not allow you to cut sections out of the middle of an audio file, and in this case you have to use a digital audio editor instead. The reason for adjusting the beginning and ending points is to remove extraneous noise and/or unwanted sections from the audio file. You may have recorded a number of different sound sources into a single audio file, and you need to separate them before the sampler can properly use them. If there is silence at the beginning of the audio file, it can be removed by adjusting the beginning of the file. You can also adjust the file's beginning past the point where the sound starts as a means to create an interesting effect. This removes the beginning of the sound while leaving the rest.

1.3.3 Playback direction

Basic editing also includes adjusting the playback direction of the audio file, which can be switched between forward and backward. The default direction is forward and results in the audio file being played as it was recorded. The audio file can also be played backward, which results in a completely different sound. This is useful as a special effect and is used less often.

1.3.4 Level control

Setting the volume level of each audio file is another primary editing function. Samplers can increase or decrease each audio file's level. If the level is turned up too high, it will result in clipping and distortion. There are other methods of changing the levels of the individual audio files, and these are discussed in Chapter 5, Section 5.4, on adding effects and processors.

1.3.5 Pitch shifting

Samplers are able to shift the pitch of each audio file. In the past, all samplers needed this function because samplers weren't powerful enough to handle having individual samples for each possible note. Instead, a single sample would be transposed to cover multiple notes using a pitch shifter. This type of pitch shifting is often very crude, with no substantial options. Modern pitch shifters, however, take into consideration what type of material is being shifted. For example, different algorithms can be used for monophonic, polyphonic, or rhythmic sources. Most samplers still use a single algorithm for all audio files. Additionally, the pitch shifting often follows an analog pitch shifting system, where the pitch is raised by speeding the sample up or lowered by slowing it down. This is a limiting factor, because large shifts become very obvious due to the change in the length of the sample. One type of pitch shifting that most samplers cannot perform is pitch correction. Pitch correctors are able to correct the pitch of the parts of sound that drift out of tune. Samplers have the ability to shift the entire sample up and down, but they can't shift a small part of the file separately.

1.3.6 Looping

Looping consists of repeating a part or all of an individual audio file. Looping is a function that has been used in the past to compensate for the technical shortcomings of samplers. Large sampled instruments require more sampler memory and an efficient method of reading the files from the storage location. Looping is a useful tool that allows sampled instruments to imitate sources that have sustained sounds. A looped section of a sample means that shorter audio files can be used to create longer playback times. Sound sources that normally have finite lengths can be sustained indefinitely when using loops. The looped section simply continues repeating as long as the key is pressed.

Looping is still commonly used. Some samplers do not need to use looped samples and can use full-length source files without worrying about memory usage and power. In situations where looping is not used, the notes can only be held out as long as the original sample length. You can make the notes shorter by releasing early, but you cannot make the note longer without using the looping feature.

Loops are still in use by samplers that are less efficient in terms of file recall and memory usage. Looping can also be used as an effect. The looped section, when repeated, might give the sample an interesting rhythm, or a loop might be used to sustain a sound that normally can't be sustained. No matter what the use, loops are still a fundamental part of samplers and a feature that is still relevant.

1.3.7 Capable GUI

Another sampler feature that has become standard is a sleek graphic user interface (GUI). This trend did not start with samplers but developed with computer technology in general. Consumers expect a user-friendly graphic interface. Many consumers associate how an interface looks with how it works; the more polished the interface, the more people assume that it works and sounds better than lesser interfaces. Although the look of a sampler has nothing to do with how it sounds, most polished interfaces do bring advancements in functionality. The most significant of these is drag-and-drop functionality. The ability to move files and effects around by clicking on and dragging a graphical representation is not a necessary function, but it can make editing and creating a sampled instrument much more efficient. Modern software samplers that are hosted inside DAWs have fully integrated interfaces. Using the mouse, you can drag audio files from the DAW's audio tracks straight into the sampler. This is a huge advantage over previous systems, where an audio file recorded into a DAW had to be exported before it could be imported into the sampler. Simply stated, what used to be a multistep process has been made into a single drag-and-drop process. When working with hundreds of files, this can make a huge difference. It has never been easier to map samples into a sampler.

1.3.8 Project manager

Along with the GUI, a file browser/project manager is a common expectation. Not all samplers have capable project managers, but if you plan to be working with a large number of audio files, this is a feature that you should consider a priority. A project manager doesn't mean a simple list of the files that are a part of each project; it includes a hierarchical display of all files, search features, file information lists, and metadata storage and recall to help with the management of the files in a sampler. For simpler sampling projects, a complex file manager might not be useful, because you have far fewer files to work with. But if the number of files you will be dealing with is quite large, then it is highly recommended that you have a fully functional and feature-rich manager.

1.4 Other expectations

There are other expectations that are sampler type–dependent. Hardware samplers have different features than software samplers. Hardware samplers come with a variety of memory amounts and different input/output configurations. Software samplers have minimum and recommended hardware configurations that determine which computers the software can be used on. In either case, you should realize that not all samplers have the same capabilities.

1.4.1 Sampler memory

Hardware sample players use wave ROM (read only memory) and sample memory. These two terms, *wave ROM* and *sample memory*, are commonly used to describe the two types of storage on a hardware sampler. The wave ROM is the permanent sound bank in the sampler, whereas the sample memory is the temporary storage area in the sampler. Sample memory uses random access memory (RAM). For readers who want to learn more about the difference between ROM, RAM, and other storage mediums (hard drive/flash), see the website for detailed explanations and a list of additional resources.

Memory types and comparisons

The wave ROM specification indicates the amount of permanent onboard storage. This is directly related to the number of samples included with the sampler. It is typical for a hardware module to have a default wave ROM size of up to 1 GB (gigabyte). The larger size means more instruments or higher-quality instruments, though usually not both. Most hardware samplers have much less than 1 GB but can accept expansion cards with additional samples. The sample memory is used to load user samples from third-party expansion sets or when recording samples directly into the sampler. Software samplers do not deal directly with the sample memory or the wave ROM; they utilize the memory and hard drive space from the computer on which the sampler is loaded. This means that the available space is much more expandable and accessible, but completely dependent on the computer system you are using. You can easily fill an entire hard drive by storing samples, and so it is advisable to dedicate a drive for this purpose. It is also not unheard of for a software sampler to come with many gigabytes of samples. Keep in mind that while it is possible to store your sample files on your computer's primary drive (that is, on the same drive as the operating system), it is recommended that a separate drive be used. This is explained later in Section 1.5.3 on disk streaming.

Expansion set compatibility chart

1.4.2 Voice count

The number of samples that a sampler can play simultaneously is limited. The term *polyphony* is used to describe the number of samples that a sampler can play at once and is measured using a unit called *voices*. The maximum number of voices for a sampler can be a limiting factor, but most samplers allow up to 128 voices. If you are using a piano-style keyboard to control the sampler, then you are physically limited to playing as many notes as you have fingers. With the use of a sustain pedal, the number of voices can be much higher, because when the pedal is pressed, the notes sustain and continue to use a voice until the sample is released. Be aware that voices can get used up fairly quickly in such circumstances. Playing a sampled instrument that has a long sustain is effectively the same as using a sustain pedal. In both cases, you can run out of voices while playing a fast series of notes. For every sample that the sampler is playing, there must be a free voice.

Some samplers have adjustable voice settings that can be switched to save processing power consumption. Adjusting the number of voices that a sampler can use is only necessary when there is a problem. For example, if some notes are not playing when a key is pressed, you know that the number of voices is set too low. Likewise, when notes that are sustained over time unexpectedly stop playing when new notes are started, you know you need to increase the number of available voices. You might consider setting the number of available voices to its highest setting, but this can affect the performance of the sampler. You could experience clicks, pops, and other audible problems when the voice setting is too high. The key is to find a good balance between too few and too many voices. Samplers often offer a choice on how to deal with too few available voices. The standard option includes cycling the samples so that when the limit is reached, the oldest playing note will release to make a voice available for the new note. Another common option is to keep a new note from starting until a voice becomes available.

1.4.3 Advanced volume control

There are several techniques used to control volume in samplers. Each individual audio file has a volume control. The volume of the audio file is determined by how hard a key is pressed. Other sampler functions, such as continuous controllers and low-frequency oscillators (both are defined below), can also affect the volume in samplers. These can control individual audio files as well as the sampler's overall output volume.

Each audio file has a maximum volume level. Quite often, this is the level at which the sound source was recorded and does not need to be changed. However, samplers can alter this level through the use of volume controls. Furthermore, when a note is played, the resulting output volume is a percentage of the maximum volume that correlates to the pressing of a key on the keyboard. The faster a key is pressed, the closer to the maximum volume the sample will play.

Continuous controllers can be used to control volume parameters. Continuous control data is a part of the MIDI protocol. It is a type of MIDI message that is used to control a variety of parameters. One source of continuous controller data is from physical controllers such as modulation wheels, joysticks, foot pedals, breath controllers, or anything that can send out a continuously variable MIDI signal. Breath controllers are used to continuously control the volume and timbre of a sample and have the potential to create very realistic performances. They are small devices that look like a whistle or kazoo, and they measure the pressure of the performer's breath. Keyboards with after-touch keep track of the pressure on the keys even after the initial key has been pressed. This continuous controller allows a keyboardist to add vibrato by applying pressure on the keys while holding notes. Another source of continuous controller data could be a MIDI sequencer. Any format that can record and playback MIDI data can also usually record and playback continuous controller data.

Envelopes and low-frequency oscillators (LFOs) are additional sampler features that can control the volume in a sampler. Most often, these affect the overall sampler volume and are not applied to individual samples, but they can be used for both. An *envelope* is a timed volume map that determines the level of a sample during its playback (Figure 1.6). An envelope (often referred to as ADSR) consists of the initial Attack, Decay, Sustain, and Release parts of the volume map. If an attack time is set to zero, then the triggered audio file will begin to play instantaneously at full volume. Settings above zero result in a fade-in. Higher settings result in longer fades. The decay controls the timing of the transition from

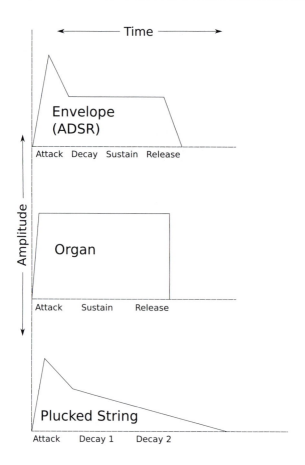

Figure 1.6 Possible Envelopes.

the full level after the attack to the sustain level. After the attack and decay are completed, the sampler holds at the volume determined by the sustain parameter. If this is set to 100 percent, then the level from the attack and decay will continue without changing. The final release determines the fade out after the sample trigger is released. When discussing ADSR, keep in mind that the envelope affects the original audio file in relation to its own volume. This means that a short audio file that has a loud attack and then quickly fades into silence needs a short envelope attack time setting to be heard. If the attack time is set to a very high setting, the audio file might be finished playing before the volume setting is loud enough to hear it. The same is true for a long release time. If the file is finished before the trigger is released, then the release time affects nothing.

A *low-frequency oscillator* can also be used to set volume changes by using an oscillator to control the volume for tremolo or to control pitch for vibrato level. An LFO will typically have different wave shapes from which to choose, and the shape of the waveform chosen will determine the resulting parameter control. If the LFO is a sine wave, then the volume will pulse smoothly and evenly, creating a tremolo style effect. If that control is used to modulate pitch, then a vibrato is created.

1.4.4 Pitch control

The fundamental pitch of a sample can be adjusted using methods similar to those for volume control (see Section 1.4.3 above). Once a sample is triggered, it is then possible for the pitch to be altered. Continuous controllers and LFOs are commonly used to adjust the pitch of a sample to create vibrato, or as way to detune a sample for creative purposes. When emulating a guitar sound, a continuous controller such as a pitch wheel might be used to create realistic string bends.

1.4.5 Spectral control

The spectral content of both individual samples and the overall output of the sampler are often altered by filters that cut and boost selectable frequencies. Continuous controllers and LFOs can control filters. Samplers also might have access to a number of spectrum effects either through built-in effects or through their host DAW. These specific effects are covered in Section 5.4 in Chapter 5.

1.4.6 Velocity layers

Since the very first samplers were invented, one goal has been to re-create realistic sampled instruments. Velocity layer switching is an important tool that helps reach this goal. A sampled instrument can have a single velocity layer or multiple velocity layers assigned to each sample. If there is only one layer assigned, then each file will be triggered over the whole range of possible velocities. If you play a single note on a MIDI keyboard, no matter how hard or soft you play, it will trigger the same audio file. Of course, the audio file will play back at a variety of levels based on the velocity of the trigger, but the file remains the same. If there are multiple velocity layers assigned, then there will be multiple files that will play instead of a single file. A sampled instrument would have separate audio files for different attacks when triggered at higher and lower velocities. The idea is to record a sound source at corresponding levels. If the source is a guitar, then the two layers might be a soft strum and a hard strum. When the files are mapped, they are assigned different velocity layers. When triggered, a lower velocity triggers the soft strum and a higher velocity triggers the hard strum. This technique can be used quite effectively to create organic instruments with realistic dynamic responses. This does not mean that every sampled instrument that utilizes velocity layers is attempting to re-create the sound source's exact dynamic range and response. Either way, every sampler should be capable of multiple velocity layers, but modern samplers are taking this idea to new levels with round-robin switching, key switching, and scripting.

1.4.7 Round-robin

Round-robin switching is a simple technique that helps prevent samplers from sounding too robotic. Any time a single sample is triggered multiple times in a row, there is the chance that it will be noticeable. Instead of playing the same sample for a given trigger, the round-robin function allows alternate sample files to be used to create a sense of randomness. This alternation can cycle through two or more possible samples. The files are usually played, one after another, in the order that they are imported. Sampled instruments that are already using a lot of different velocity layers might be bogged down by also adding round-robin files to every single note. In this case, using either round-robin files or a lot of velocity layers, you'll be able to find a highly realistic and organic result.

1.4.8 Key switching

Key switching is an exciting tool for live musicians or for those looking for ways to be more efficient in sequencer programming. Key switching is a way to load several different sample sets at the same time, and it allows you to switch between them by pressing a single key or using a continuous controller. Imagine an instrument like a violin that is capable of many different bowings. You can use a bow with several different styles of playing, but you can also use your fingers to pluck the strings. In the past, each of these quite different styles was encased in a completely separate sampled instrument, but now in some samplers they can be combined to form a single patch that can use key switching to change between each of the various playing styles. Sometimes you can also use a continuous controller, such as a modulation wheel, to scroll through the different styles. You can imagine that this process relies heavily on the power of the sampler or computer system. Only a few modern samplers have this feature, and these samplers have systems in place to help with the efficient use of the key switching technique.

1.4.9 Scripting

A recent trend in samplers is the ability to use a programming language to add functionality to the sampled instruments. A company called Native Instruments was the first to use the term *scripting* in the context of samplers and was the first company to create a sampler that offers programming options to the end user. While scripting is not an easy skill, it has been used with great effect in a number of sampled instrument releases. The advantages of scripting are most obvious when creating hyper-realistic sampled instruments. You can create an instrument that interprets the control information from whatever controller you are using and then converts it into a realistic performance that could exist on the replicated instrument.

For instance, if you are using a piano-style keyboard to play a sampled trumpet, then there are some fundamental differences between the instruments. For starters, a piano can play many notes at once whereas a trumpet cannot. The script would translate the input data, and the trumpet would stick to realistic parameters. Samplers can already be set to play only one note at a time, and so the scripting feature might seem redundant in this case, but it does much more. You can script the trumpet to react to different keyboard patterns in different ways. If two notes are played very close together, then the sampler might play the trumpet slurring from one note to another. If you play two alternating notes, then the sampler might play a trill sample. You can script dynamics into the instrument; a trumpet player will run out of air eventually, and so you can script in separate samples for when there are no breaks, and the sampled instrument will actually sound like it is running out of air. With scripting, there is a lot of potential for creating realistic sampled instruments.

1.5 Additional software sampler features

Software samplers are pushing the current computer limitations in several key areas. Using a sampler to play large sampled instruments can tax even the most powerful system, and in most cases software samplers are designed to be used as plug-ins with DAWs. Plug-ins can be used multiple times in a session, and for every instance of the sampler you will need the necessary processing power. Features have been introduced in some samplers to help with this issue: low-resolution playback, hold all in memory function, and streaming. The primary goal of each of these features is to help with loading the files from the sample storage hard drive. Drive speeds are often a limiting factor when determining the number of sampler plug-ins that can be used in a given session.

1.5.1 *Low-resolution playback*

The low-resolution playback option allows the files to be used at a lower resolution to help minimize the amount of computer resources required for playback. Lower-resolution files take up less space in the computer's memory, and so they use less of the computer's resources. This is nice in situations where quality is not a top priority. If you are writing a song, then having the absolutely highest quality might not be very important, and using a low-resolution play option will free up resources for other tasks. But, if you are creating something that requires the highest quality, then you should avoid the low-resolution playback option. In this case, you might be able to get away with using the low-resolution playback up until the point that you complete the final mix down. The specific implementation of this varies from sampler to sampler.

1.5.2 *Hold samples in memory*

The ability to hold all of the samples in the memory of your computer removes the workload on the hard drive except for the initial instrument loading. The downside to this feature includes (1) using memory resources that are needed by other functions, and (2) not having enough memory to load multiple samplers simultaneously. Current computers can have many gigabytes of random access memory (RAM), and this might eventually be the ultimate solution to slower hard drives.

Other storage alternatives

1.5.3 *Sample streaming*

Streaming is one of the more powerful solutions to limited resources. *Streaming* uses a combination of RAM storage and hard drive storage to efficiently load sampled instruments for playback. A small section of the beginning of each file is stored in RAM, and when a note is triggered, the sampler plays back the beginning of the file and reads the rest from the hard drive. This allows a more efficient sample loading before playback, and it also provides time for the sampler to load the rest of the files without taking up too much RAM. Sample streaming allows samplers to use long audio files without being limited by RAM.

1.6 Benefits to using samplers without creating your own instruments

It is possible to take advantage of the power of samplers without ever creating your own instruments. Even if you would like to create an instrument at some point, you can still use the sampler's default sounds and available expansion packs. Keep in mind that most people who create sampled instruments also use samples made by other people. There are good reasons to forgo the sampling process and capitalize on the efforts of individuals and companies who create samples.

The reasons to use other people's samples instead of creating them yourself include achieving immediate results, accessing hard-to-find sound sources, and taking advantage of an individual's or company's professional experience in sample creation. Creating sampled instruments is not usually a short process. Creating a complex instrument can take many hours of planning, recording, editing, mapping, and tweaking. Many projects will use multiple sampled instruments, and you will not likely have the time to create every instrument yourself; it just isn't a realistic expectation. Along with the time requirements, you might not even have access to the sound source that you wish to re-create with a sampler. If you want to create a sampled Stradivarius, then you need a Stradivarius. Since a Stradivarius violin can be worth millions of dollars, it is unlikely that most sampler enthusiasts would have access to one. Lastly, there are professionals who create sample banks for a living. It is possible for you to create a professional-quality instrument, but mastering the skills required will take time and personal investment. If you need an instrument before you have reached that skill level, then using someone else's instrument shouldn't be an issue.

1.7 Using samplers

Samplers are used in a variety of settings and for a variety of reasons. The general categories include music preproduction, production, live performance, composition, and audio for video postproduction.

Definitions and examples

1.7.1 Preproduction

In the music preproduction phase, samplers are a great tool that can help with songwriting and demo creation. During the orchestration process, a single person can have access to every instrument imaginable. Instead of calling in a musician to record different parts for a song, only to rewrite the part later, a songwriter can use a sampled instrument to create the parts that can be changed as often as desired. The parts can also be tested on different instruments with the click of a mouse. Of course, permanent part changes will need fine-tuned adjustments to compensate for different performance options and techniques on different instruments. The orchestration process can transition seamlessly into the creation of a demo. Samplers sound good enough to be used as a demo of a song for publishers and producers. In some musical genres, samplers are used in the next phase of record production as well.

Definitions and examples

1.7.2 Production

During the music production phase, samplers can be a very useful tool. Some sampled instruments sound as good as or better than the alternatives. It would be nice to have the best studio musicians at your fingertips, but this is not always practical, and some projects require something that studio musicians might not offer. Ukulele virtuoso and widely recorded studio bassist Lyle Ritz released a solo ukulele and bass album in 2006 called "No Frills." While most would have guessed that Lyle played the bass parts on his upright

27

bass, he chose to use a blend of two bass samples. Having played on thousands of well-known records, you might wonder why he used a sampled instrument instead. When the album is heard, it is very hard to tell that the bass is not a real upright bass because it is programmed so well.

Many genres of music require the use of samplers because the sounds that are expected are not real instruments. Highly processed drum sounds and stylized piano sounds are two examples of instruments that samplers can make accessible during the production phase. Samplers are also achieving highly realistic sounds that are taking over the roles of real musicians. Sarah McLaughlin's holiday album "Wintersong" utilized string samples instead of real string players. Many listeners will never know that the strings sounds come from someone playing a keyboard instead of a group of musicians playing actual stringed instruments.

Definitions and examples

1.7.3 Live performance

Samplers have always offered musicians a great alternative to having a lot of instruments on stage during live performances. A solo musician can learn how to play a single controller and then will have access to hundreds of different sounds. While this is an oversimplified view of using sampled instruments on stage, it really can save a lot of resources: it can protect valuable instruments from traveling on tours, free up money that would have been spent on specialized instrumentalists, and save on stage space because a couple of rack spaces will hold all of the instruments.

Definitions and examples

1.7.4 Composition

Composers have long pushed the development of sampling tools because they can help realize many different sounds and techniques that are not easily possible through non-sampler techniques. Composers often push the creative abilities of music technology in ways that popular music production does not. (The popular uses more often than not attempt to re-create realistic instruments.) Composers can use samplers to help hear their music before it is played by traditional instruments, and samplers can also be used to create new and innovative sound landscapes that no one has ever heard before.

Definitions and examples

1.7.5 Postproduction

During the audio for a film or video postproduction phase, samplers can be used to great effect for demo creation, orchestration, and during final production. Samplers can be used to help film composers create their scores and are often used as a part of the final music track. Many scores blend traditional music with pop-style music, and samplers are an efficient tool that can assist in this process. Samplers can also be used in the SFX and Foley creation process. Imagine having instruments that have a variety of different SFX such as footsteps. When lining these up to film, a performer could actually play the footsteps using a MIDI controller instead of requiring a Foley artist. This holds true for all SFX and Foley.

1.8 Benefits to creating your own sampled instruments

1.8.1 Low-cost solution

There are some benefits to creating your own sampled instruments. This section outlines some of those advantages. The first of these is saving money. Buying large expansion packs might seem like a good investment, but if you can create some of the instruments yourself, you could save yourself a few hundred dollars. It really comes down to investing your time in the project; if you have the time and technical resources, then it can save you money.

1.8.2 Need-based solution

Another reason to create your own sampled instrument is to create specific need-based solutions. If you would like a single instrument, then certainly you can go out and buy a sample library that has the instrument. But you are also buying a lot of extra and possibly never-to-be-used instruments. This can be a waste of money. Instead, you can create the single instrument you need and save it for future use.

1.8.3 Unavailable instrument

Sometimes you may not be able to find the instrument you are looking for, or the one you find might not sound like you want it to. This is a valid reason to create your own sampled instrument. Sample libraries do not have every possible instrument. Also, if you have a homemade instrument that is not available anywhere else, you could create a sampled version of the instrument for use in different settings. Sampling is an extremely flexible process that puts you in control of the end product.

1.8.4 Instrument preservation

Creating a sampled instrument is an excellent way to preserve the sound of an older or "dying" instrument. Instead of continuing to play the instrument, you can use it to make a sampled instrument that can be played as often as desired with no negative side effects. Taking certain instruments on tour might not be ideal on account of undesirable wear and tear, and creating a sampled version will prevent any unnecessary deterioration.

1.9 Summary

The process of creating a sampled instrument can be time consuming, and it often requires a lot of planning and effort. Although it is a lot of work, creating your own samples brings personal satisfaction and gives you a creative outlet. Creating a high-quality sampled instrument is not a small project. Most of the time you will be doing it for no money, with only your personal projects in mind. This doesn't mean that creating your own instrument isn't worth it.

If you are able to create many different instruments that are high quality and have interesting sounds, you might consider trying to sell them to companies that specialize in offering sample libraries. There is nothing wrong with turning a hobby into a profitable situation. With the number of samplers that are available, there is always a need for additional fresh sample libraries to be released.

Samplers are powerful tools that are both flexible and accessible. Whether you are using a sample player or creating your own sampled instruments, you should now be aware of the primary features offered in most samplers. Throughout the rest of this book, you will learn how to use specific sampler features to create sampled instruments. Whether you are creating a complex instrument that takes many days to complete or a simple instrument that takes only minutes, these features are the tools you will use.

Throughout this chapter, we have looked at a lot of features that modern samplers use. Most of these features have been introduced but have not been explained in great detail. As you continue through the rest of this book, we will build on this introduction by explaining how to create a sampled instrument. The approach is practical, and you are often expected to glean the additional explanations by reading between the lines. If you need further and more up-to-date information, be sure to check the website.

MAKING
CONNECTIONS 2

The focus of this chapter is to help you prepare for the sample recording phase of the sampling process. This preparation involves ensuring that you have the appropriate equipment and that it is hooked up correctly.

2.1 Signal chain

The signal chain is critical to understand. It can be quite complex, and it's important to have a good grasp on what items you need and how to hook them all up. At any given moment, you should know where your audio data is. An example of the order of equipment for a microphone-based system is microphone—preamp—analog-to-digital converter—audio interface—computer—software—playback system (see Figures 2.1 and 2.2).

Each of these components is described later in this chapter. While this chain describes one of the most extensive possibilities, there are many different variations. A hardware sampler, for example, often has the simplest chain in terms of actual equipment connections, with a microphone being the only piece that needs to be plugged in. Just because the chain seems simple, however, it does not mean that there is no preamp or converter. Rather, the hardware sampler has all the components built into it. That is the case for many different computer audio interfaces: the interfaces usually have preamps and converters built in.

There are alternatives to using the built-in components. You can use separate components, which lets you customize the signal chain any way you like. There are very nice micro-

Microphone - Preamp - ADC - Hardware Sampler

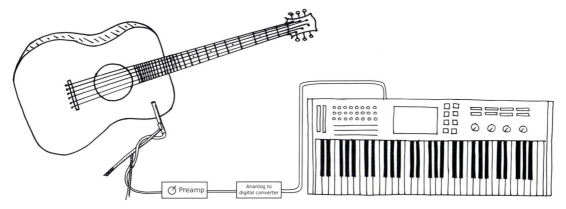

Figure 2.1 Signal Chain 1.

Microphone - Audio Interface (preamp & ADC) - Computer

Figure 2.2 Signal Chain 2.

phone preamplifiers that can only be purchased separately. If you want a higher-quality analog-to-digital converter, you can buy one and use it instead of the built-in converter. These components can be used with hardware and software samplers to raise the overall quality of your signal chain. But there are also benefits to using the built-in components, one of which is simplicity. It is simpler to hook up and use. It is also cheaper, because you don't have to purchase any external components.

2.2 Capturing the sound source

Before we discuss hooking things up and troubleshooting, let's cover each of the basic capturing tools in greater detail. To create a sampled instrument, you need a sound source and something to capture the sound source. There are two primary methods to capture the source: one involves a microphone and the other involves a direct input method. A microphone is a transducer that converts acoustic energy into electrical energy. The direct method accepts sources that are already electrical and does not convert the signal. The direct method might be easier in terms of recording, because you can connect the source directly to the sampler with one or two cables (one for mono and up to two for stereo), and then you can immediately record. For other sources, you need to set up a microphone and capture the acoustic sound from the source. The method you use depends on the sound source. Some sources cannot be recorded by a direct method because they do not have an electrical output. For sources such as acoustic guitars, if they also have magnetic pickups, you can capture using either a microphone or a direct method.

For an excellent book on microphones, see John Eargle's *The Microphone Book* (2nd edition, Focal Press, 2004). It is amazing that nearly 400 pages can be written on a single subject such as microphones. This gives you an idea of just how complex the topic really is. Our discussion here takes a much simpler approach. The basic types of microphones are dynamic, condenser, and ribbon. Each type works in a different way, with different strengths and weaknesses. You should take your project goals into consideration when choosing which microphone to use. While it is possible to pay thousands of dollars for a single microphone, you may be investing too much; a hundred dollars may buy something that would work just as well or better for your specific application.

Instead of describing each type of microphone in detail, this section includes some basic information that will help when using a microphone to record your samples. As many have learned from experience, it is difficult to figure out in advance what the most suitable microphone for your use may be. Regardless of whether you use only the microphones you own or whether you rent very expensive microphones, until you hook them up and listen to the results, you cannot be sure what will work best. You might be able to make an educated guess because the different types of microphones have specific traits that are consistent, but sometimes it turns out differently than you expected.

2.3 Using microphones

2.3.1 Power

Microphone sound samples

When hooking up a microphone, you need a microphone, a cable, a preamplifier, an analog-to-digital converter, and either a hardware sampler or an audio interface for your computer. If you are using a condenser microphone, you need to have a power source (Figure 2.3). This power can range from +1.5 volts DC to +48 volts DC (direct current). Electret condenser microphones may have power supplied by a small battery built into the microphone. It is essential that you follow the specifications for the voltage and polarity of these batteries. If the voltage for the microphone circuit is supplied by the power supply of your recording console, it is called *phantom power*. The phantom power voltage is distributed over two of the conductors of your basic balanced microphone cable (see Sections 2.10 and 2.11 on audio cables later in this chapter) with the audio signal. This DC power does not interfere with the AC (alternating current) of the audio signal. Not all professional microphones use the standard +48 VDC power specification. The proper power voltage is normally identified by the manufacturer. Follow the specification of the manufacturer to ensure there is no damage

Figure 2.3 Tube Microphone w/Power Supply.

Tube Microphone w/ Power Supply

Output to Preamp   Polar Pattern Switch

caused to your microphones. Do not apply phantom power to ribbon microphones. The delicate ribbon material can be damaged.

2.3.2 High pass filter

Many microphones have additional features that provide extra value and flexibility. These include a high pass filter, a pad, and a variety of polar patterns. A high pass filter allows high frequencies to pass while attenuating low frequencies (Figure 2.4). This is useful for eliminating unrelated low-frequency noise, reducing the sound caused by vibrations and bumping the microphone stand, and reducing the proximity effect. The *proximity effect* is a boost of lower frequencies and exists in situations where a sound source is close to a directional microphone. The result is a boost in the amount of bass.

2.3.3 Pad

A pad lowers the electrical output of the microphone (Figure 2.5). This can be useful when capturing loud sounds. A pad might be used while recording drums, electric guitar ampli-

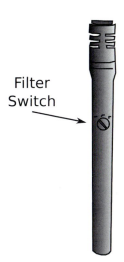

Figure 2.4 Filter Switch.

Filter
Switch

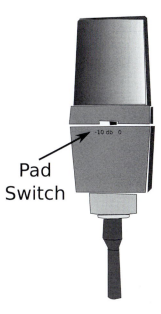

Figure 2.5 Pad Switch.

-10 db 0

Pad
Switch

fiers, and other loud sound sources. Using a pad is very important when the sound source is loud enough to cause clipping and distortion at the microphone's output. With the pad engaged, any clipping and distortion will be avoided. Keep in mind that certain sounds can still be too loud and might cause clipping before the pad phase of the microphone. In this case, lower the volume of the sound source or increase the distance between the source and the microphone. Many preamplifiers and audio consoles also have pad switches.

2.3.4 Polar patterns

A polar pattern describes the directionality of a microphone (Figure 2.6). There are three primary polar patterns and several others that are related to the first three. The primary patterns are omnidirectional, bidirectional, and unidirectional. The other types (hypercardioid and supercardioid) are subcategories of the unidirectional pattern. The omnidirectional pattern captures a pressure reading of a specific point in the air without regard to the direction from which the sound comes. A bidirectional pattern captures a differential pressure reading using a diaphragm that is open on two sides. It receives sound from two directions and has reduced reception coming from the sides. The unidirectional pattern captures a cardioid (heart-shaped) pressure reading from the front of the microphone. A unidirectional microphone can either be the result of combining an omnidirectional pattern with a bidirectional (figure-eight) pattern or can be created through the use of specific design features of the microphone's encasement (see Eargle's book listed in Section 2.2 for more

Figure 2.6 Polar Patterns.

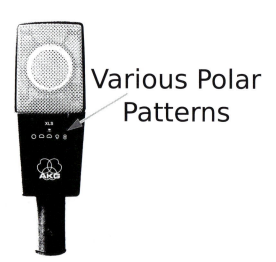

on this). The other unidirectional subcategories are created either through different omnidirectional and bidirectional combinations or through a variety of different physical designs.

2.4 Using a line input/instrument input

Instead of using a microphone to capture your sound source, you can sometimes use a direct input, such as a line input or an instrument input. Although the actual input on either a hardware sampler or on a preamplifier for line and instrument inputs might look the same, they are different. The primary sound sources you might capture using these methods include synthesizers, keyboards, electric guitars, bass guitars, or any source that comes from a computer or other hardware sampler. Guitars, bass guitars, keyboards, and synthesizers are typically recorded through an instrument input, but some synthesizers, keyboards, and all computers/samplers will utilize the line input. The difference is in the electronic impedance common to each of these sources.

There are three common electrical signal levels used in audio electronics. These are colloquially referred to as *microphone level*, *instrument level*, and *line level*. Each signal varies in strength and results in different electrical impedance requirements. The impedances for each type vary quite a bit. An electric guitar might feed into an impedance of 1,000,000 ohms (1M). A low-impedance microphone might feed into 250 ohms. A line level input on a console might accept around 10,000 to 50,000 ohms (10–50 k). Without going into much detail, this information is designed to reinforce the differences between each of the three signal levels. A microphone preamp amplifies a microphone signal many times to bring it to line level. A preamp that accepts instrument input levels also amplifies the instrument signal to line level. Mixing consoles and other audio equipment operate at line level, so the objective is to use appropriate preamplification to bring all sources into the signal path at this established line level.

Figure 2.7 Direct Box.

Most hardware samplers have either a microphone preamplifier or a line-level input. Some have instrument inputs as well. If you want to connect an instrument-level output to a sampler but the interface has only a microphone input, then you can use a direct box, or DI (direct injection). A direct box is a transformer that takes a line-level signal and steps it down closer to a microphone-level signal (Figure 2.7). You can then hook the direct box into the microphone preamplifier. A direct box also converts an unbalanced signal to a balanced signal. Unbalanced and balanced signals are discussed later in this chapter (see Section 2.11).

If you are using a software sampler in your computer, then the input choices depend on the specific audio interface. Some audio interfaces, especially those designed with a non-professional audience in mind, have a variety of different input options. If you operate in a home studio, then it is great to have a single interface that has line input, instrument input, and microphone input. It is also nice when using a portable interface because it is more efficient to have a single piece of gear.

2.5 Using microphone preamplifiers

Microphone preamps are a tricky part of the audio chain. The only way you can really tell if you have one that is good (or bad) is to compare it to other preamps, although sometimes you can detect obvious flaws. Some people claim they can identify the traits of various preamps and hear how preamps "color" the sound differently. Still, when capturing a variety of sound sources using a variety of microphones, it can be difficult to predict the success of the preamp. Recording studios often like to have various preamps on hand because the engineers feel that certain sound sources require different microphones and preamps.

A preamp, as stated earlier, amplifies the microphone signal to a line-level signal. This is required for standard processing, including preparation for conversion from analog to digital. The electronics used in the preamp determine the sound of the preamp. Some preamps use solid-state amplification; others use tubes for amplification. Some preamps use transformers while others do not. You may well be satisfied to use the preamp built into your sampler or your audio interface. But if you choose to purchase even one additional preamp, it can be hard to decide which one to buy. Preamps come in a wide range of prices and styles. You can easily spend thousands of dollars for a single preamp. The best approach is to try out as many as possible. Don't simply buy the one you can afford. Sometimes a specific preamp style just isn't right for the recording you wish to do. Sometimes the most expensive preamp isn't what you are looking for. You might consider buying a very clean and transparent preamp for general use, because such a preamp often works well on a variety of sound sources.

When testing out potential preamps, here are some questions to ask yourself:

1. Does the preamp change how the natural source sounds? How?
2. Can the gain of the preamp be pushed very high without a lot of distortion?

3. What other features does the preamp have? (equalizer, compression, digital outputs)
4. How expensive is the preamp?

When testing out the preamp, use a single microphone so you can make accurate comparisons. Consider using a measurement microphone, such as the Earthworks M30, which has a flat frequency response. This helps the preamp quality and character shine through.

2.6   Using analog-to-digital converters

Once your signal is inserted into a preamp from a microphone, an instrument input, or a line-level input, you have to take the analog signal and convert it to digital. Hardware samplers and audio interfaces have analog-to-digital converters built in. The question is not whether you have access to an analog-to-digital converter, but how good it is. Some external preamps have great analog-to-digital converters, but you can also find stand-alone converters that do nothing but analog-to-digital conversion. In listening tests, it is possible to hear the difference between poor and great converters, but the real challenge is not in the choice of converters but in how to hook them together.

With a hardware sampler, you can only use an external analog-to-digital converter if there are digital inputs. The type of digital input is important because there are several different digital formats and they are not compatible with one another. The most commonly used digital formats are S/PDIF, ADAT optical, and AES3 (these are discussed later in this chapter; see Sections 2.9 and 2.10). It is important to understand the different types of digital signals.

Likewise, it is critical to understand how a word clock functions in a digital audio system. Every audio system that deals with digital information needs to have a master clock that regulates the movement of all digital information and synchronization between different pieces of equipment. If you have more than one master clock, there can be "clicks and pops" in your audio signal. Sometimes the equipment won't even play audio.

The master clock can either be the internal clock of one piece of equipment (every piece of equipment that uses digital signals has one), or it can be an external box designed to be a master clock. To connect two clocks together without an external master clock, you can use the digital cable you are already using. AES3, S/PDIF, and most other digital signals have clocking signals embedded with the audio data. You only need to tell the equipment who is the master and who is the slave, and then it will work. This is sometimes a menu choice or a switch on the back. See the equipment's specific instructions for more information.

The other method for connecting a master clock to a slave is through word clock input and output ports. These typically use BNC connectors with 75-ohm cables. Depending on the system, you will likely run cables from the master to a slave and then from that slave to the next until all pieces are connected. The last piece of gear typically connects back to the master, creating a loop. A second option is to have a master clock with multiple outputs. Each output connects directly to each piece of equipment. In this case, each slave needs to be terminated using a 75-ohm word clock termination plug.

In addition to sending clock information, most gear has to be set to master or slave mode. While you can purchase a piece of gear that is a dedicated clock, you can also use nearly any piece of digital gear. Not every piece of digital gear has a great clock (poor ones can create timing errors called *jitter*). If you are going to use an analog-to-digital converter instead of the hardware sampler's converter, you have to make sure that the sampler or the converter is set to be the master, with all other devices set as slaves.

One large benefit of using a separate analog-to-digital converter is that you can bypass all of the analog components on a hardware sampler. Instead of using a decent preamp and decent converter, you can get really great components and preserve the quality by sending the data into a sampler using a digital format. Digital information is able to retain a higher level of accuracy when transferring signals from one place to another, while analog transfers are more susceptible to disruptive noise and interference.

2.7 Using audio interfaces

An audio interface is a basic requirement when using a software sampler. Most computers have some sort of microphone input and speaker output, but using the built-in hardware is not a good first choice. Many onboard microphone inputs use cheap components and are very noisy. An external audio interface can be used to bypass the built-in hardware, and this creates an environment with much higher-quality possibilities. The different types of interfaces use expansion cards, USB (universal serial bus) ports, and FireWire (IEEE 1394) ports. There are also other miscellaneous connection protocols, but these three are the most common.

2.7.1 Expansion cards

Expansion cards are a standard way to connect an audio interface. Some of the most powerful audio interfaces use expansion cards that are installed inside a computer. Some expansion cards even have additional computing power available in the form of DSP (digital signal processing) chips. The DSP removes some of the processing from the computer, freeing up resources for use in other places. Expansion cards can have a variety of digital and analog inputs/outputs, but one of the best options is to use a card that provides many digital ins and outs. The quality of the audio is then reliant on the quality of the external analog-to-digital converters. This means that the expansion card acts as a gateway into the computer, but you do need separate preamps and converters. If money is not an issue, this is an ideal way to go. But there are some good expansion card interfaces that also have preamps and analog-to-digital converters.

2.7.2 USB and FireWire

USB and FireWire ports are used as a means to connect audio interfaces to computers. There are multiple USB and FireWire types, including USB 1.0, USB 2.0, FireWire 400, and FireWire 800. The latter versions of each type are the newer versions, and these can transfer larger amounts of data in less time. When choosing an audio interface, you should be aware of two issues. The first is compatibility, so you should research which interfaces are compatible with different computer hardware. Most manufacturers publish a list of equipment that has been tested with specific computers, including configurations that work and some that should be avoided. Just because your computer has a FireWire port doesn't mean that every audio interface that connects using FireWire will work. Sometimes only a certain manufacturer's FireWire port will work; likewise for USB 1.0 and 2.0. Another thing to watch out for is how an interface is powered. Some USB and FireWire interfaces can be powered directly from the computer, but if you hook too many devices to the same port, there might be insufficient power.

2.7.3 Miscellaneous connections

Other methods of connecting audio equipment to a software sampler include mLAN, digital inputs, and wireless inputs. mLAN is a proprietary format that is available in some cases but is not yet widely used. mLAN can transfer a number of different audio channels, along with other information such as MIDI data and control data. It uses a computer networking protocol when transferring data and works fairly efficiently. Another way to get audio into your computer is through any existing digital inputs on the computer. More and more computers now come with a set of digital inputs and outputs, but this can be a limiting factor because typically there is only a stereo input and output. Wireless connectivity remains relatively undeveloped due to limited wireless bandwidth. However, it will not be long before everything is connected wirelessly. Imagine a world without cables! While not unrealistic, this technology has yet to be fully developed.

2.7.4 Monitoring

The final step is the playback system (monitoring). Do not cut corners here. Most computers have poor audio playback. You will be making some very critical listening judgments during sound sampling and editing, so it is advisable to invest in a quality listening environment.

2.8 Using an alternative to audio interfaces

A sound source need not be recorded directly into a sampler. There are other methods that can be used, even if they add extra steps into the process. The primary alternative is to use a completely separate recording system and then transfer the files into the sampler. With software samplers, this is already the primary method. Most software samplers do not have recording capabilities. With software samplers, you will most often use the samplers as a plug-in to a host. This creates a very efficient workflow, because all the power of the DAW can be accessed by the sampler.

The different recording methods include having a second DAW (when using a software-based system), a CD/DVD recorder, a mini-disc recorder, a flash recorder, a hard drive–based recording system, or a video recorder format. Each of these has strengths and weaknesses. For example, an advantage of using a portable flash recorder is, obviously, portability. A flash recorder can fit in the palm of your hand. The storage medium is also quite easy to use when transferring to your sampler. Some flash recorders can be hooked directly to your computer, and the files can be copied with little or no hassle. The disadvantages are that the quality is generally questionable and there are extra steps involved.

Each method can be used with a certain amount of planning and ingenuity. This process includes a recording session to one of the above formats, followed by an audio dump into an audio editor, and then a dump into the sampler. Already you can see that there are disadvantages to doing it in this way; it is a complex process with more steps involved, especially considering that there are alternatives that work better and more efficiently.

2.9 Hooking it all together

With so many technologies available for use, it is not always easy to make the correct connections. Which cables should you use? Are they all the same? The following section is designed to help answer these questions. Things are changing so rapidly that if something is not covered here, don't feel bad about calling an expert for help; choosing the right cables and connectors can make a difference.

The first important distinction is between a cable and its connector. There are a number of different connectors that can be attached to a variety of cables. Figuring out what you need can be answered by the following questions:

1. What connector type does the first piece of gear have?
2. What connector type does the second piece of gear have?
3. Are they balanced or unbalanced?
4. Is the signal analog or digital?

One of the most common connectors you will deal with is the RCA connector (Figure 2.8). This is one of the older connectors, created by the Radio Corporation of America in the 1930s for use in early radios and phonograph players. It is also known as the *phono connector*. It has become the most widely used audio connector because it connects most consumer audio gear as well as a lot of video equipment. This is the connector that most CD/DVD players use when connecting to TVs and amplifiers. The male RCA connector has a round opening with a single pin sticking out. The female connector is just the opposite, and the connectors should fit snugly into each other for a good connection.

Figure 2.8 RCA Connector.

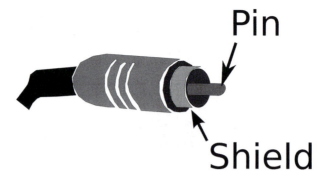

There are usually two sub-types of each kind of connector. One gets plugged into the other. To differentiate between the two, one is called the *male* and the other the *female*. The male connector is the one with the pin(s) which are plugged into the female socket. While this seems like a crude reference to human anatomy, it is the standard terminology and carries no inappropriate connotations.

The RCA connector can only carry a mono signal. It is unbalanced (see Section 2.11 below) and is used mostly with analog audio and video signals. However, RCA connectors are also used for digital audio signals. This can become confusing because, while the connectors are the same, the cables they are attached to are not. The typical analog RCA cable that comes with your CD player (with red and white connectors) is adequate for analog audio but does not work properly for digital audio. Digital signals have much higher bandwidth requirements and need different cable capacity. The most common digital signal that uses RCA-style connectors is S/PDIF (Sony Phillips Digital Interface) and requires a cable specifically designed for digital content. The S/PDIF format specifically requires a 75-ohm cable. If you are sending a digital signal over an RCA-connected cable but are uncertain if it is the right kind of cable, then it probably is not. Digital cables are usually branded with a stamp indicating that they are digital cables.

Figure 2.9 XLR Connector.

Another connector type is XLR (Figure 2.9). Cannon Technology first invented this under its X series of connectors. The L stands for *locking* and the R for *rubber insulation*. This connector normally has three pins in a balanced configuration. However, there are several other configurations with more than three pins. The reason this connector has been so popular is that it locks into place and so cannot be accidentally disconnected. This is an important feature in an industry where there is very little room for error and even less tolerance for down time.

The XLR connector is also used by both analog and digital formats, and the same issues apply to cable choices. The analog use of the XLR connector is normally associated with microphones and balanced lines. The digital audio use is normally associated with AES3 (a format standard set by the Audio Engineering Society and widely adapted for professional use). Again, it is important to get the cable that is designed for use with microphones or for AES3 because they are not interchangeable. An AES3 digital signal uses a 110-ohm cable.

Another connector is the 1/4-inch TRS/TS. This connector was developed for early use in telephone switchboards (patch bays) and is also known as the *phone connector*. The TRS (Figure 2.10) means tip-ring-sleeve and can be identified by two bands around the top of the connector, while TS (tip-sleeve) only has one band (Figure 2.11). The TRS connector can be used in different ways. One such way is as a stereo connection with T = left channel, R = right channel, and S = ground/shield. Another way is as a balanced connector with T = signal (+), R = signal (−), and S = ground/shield. The last way it is commonly used is as an insert connector with T = send, R = return, and S = ground/shield. This means that if it is inserted into a plug that accepts a TRS insert, the audio will be sent out the 1/4-inch plug into the other piece of gear and then back along the same cable into the original piece of gear. The TS configuration is most commonly a mono and unbalanced connector with T = signal and S = ground/shield.

Figure 2.10 TRS Connector.

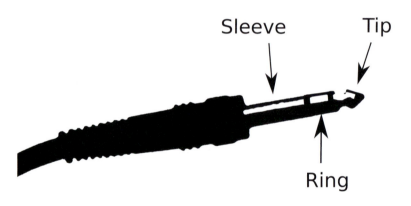

Figure 2.11 TS Connector.

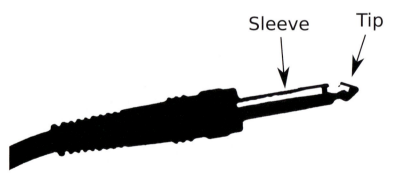

As always, the connector can be used for analog or digital. Generally, 1/4-inch connectors are not used for digital audio except in patch bays in recording studios.

Similar to the 1/4-inch connector is the 1/8-inch connector, or 3.5mm jack, which is a mini version of the 1/4-inch. It is also known as the *mini phone connector*. This connector is commonly used with headphones and computer sound cards. The same configurations are used with this mini jack as with the 1/4-inch, except that the TRS is the most common and is almost always used for stereo signals.

Another connector is the banana plug (Figure 2.12). This has one pin that sticks out that is not protected by a physical shield. It is often used as a pair of pins, as illustrated.

These connectors are used mostly for hooking speakers to amplifiers. They are usually color coded with black and red and should always be plugged in accordingly. Often, with commercial speakers and amplifiers, there are wires that go from the amplifier to the speakers without connectors; these are then attached with a screw clamp or just a clamp. There are two wires that are always color coded or labeled "+" and

Figure 2.12 Banana Plug Connector.

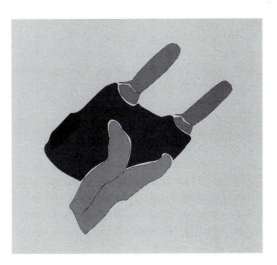

"–." The +/red cable represents the signal, and the –/black cable represents the ground. The wires must be hooked to the proper place or the polarity of the speakers will be reversed. The banana plugs are a way of creating a solid connection, one that is more reliable than the plain wire which inevitably becomes bent out of shape. Banana connectors are also used for oscilloscopes and other professional test equipment. The banana connectors are rarely used in connection with digital audio.

One last connector that can have multiple cable types is the BNC connector (Figure 2.13). BNC stands for Bayonet Neill-Concelman, named for its inventors. This is a step up from the RCA-style connector but is used in a similar way. The BNC connector is a locking connector that is commonly used with oscilloscopes, video signals, word clock signals, and other digital signals. BNC connectors are used mostly in professional applications, but many of these applications have now been adapted to work with the RCA connectors. For example, commercial DVD players that have component video (three connections which carry the RGB video signals for better color separation) now use RCA connectors instead of the more traditional BNC connectors. Professional studios

Figure 2.13 BNC Connector.

Locking
Mechanism

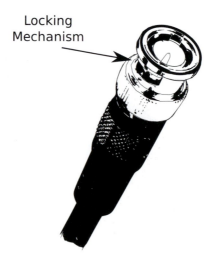

and broadcasters still use BNC connectors for the same task. BNC connectors are also used when connecting word clock signals between professional audio equipment. BNC connectors can be used for both analog and digital signals and are a good choice because they lock into place much like the XLR.

The optical cable/connector (toslink connector) that was once made popular by Alesis ADAT technology is currently a general-use cable. Optical connectors (Figure 2.14) require a specific cable, and they cannot be used with other cable varieties. The shorter versions of this cable primarily use plastic fibers to carry the optical signal; longer versions use glass fibers. This technology has allowed longer and cleaner signals to be passed through cables. In addition, several companies have created multichannel formats that can be passed through optical cables. That is important because surround-sound formats can use a single optical cable.

It is easy to identify the source and destination ports with optical cables even though both connectors look the same. Turn on the gear without the optical cables plugged in, and you will see a red light coming from the output, and the input

Figure 2.14 Optical Connector.

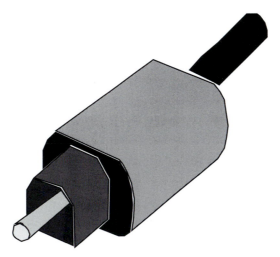

will be dark. The red light is a laser, which is how the data is transmitted and flows down the optical cable. Making a note of which port has the light and which doesn't will tell you where to plug the optical cables.

2.10   Cable types

While the connector types listed above are critical, it is also important to have a basic understanding of the different cable types. It is very rare that using the wrong cable will actually damage equipment, but it can be a detriment to the quality of your audio. There are four primary types of cables: line-level cables, microphone cables, digital cables, and speaker cables. In an effort to keep this simple, you can use line-level and microphone cables interchangeably if used over short distances. Microphone cables require a flexible shield because they move around quite a bit, whereas line-level cables are more rigid. Digital cables come in two types: 75 ohm and 110 ohm. S/PDIF uses 75-ohm cables, and AES3 uses 110-ohm cables. Speaker cables are typically made with thick copper, because they are used with amplified audio signals and so require a large current-carrying capability. An example speaker/amp rating is 8 ohms. If you are the type of person who likes to make your own cables, then you can buy everything individually and assemble the pieces. If you are not sure which cable you need, you can always buy ready-made cables. But then it is a matter of ordering the right ones.

The connectors and cables are straightforward, but grounding, clocking, and balanced/unbalanced cables are more difficult concepts. While not easy, it is worth the time to learn about these technologies because you will use them while creating samples.

2.11   Balanced vs. unbalanced

The balanced cable has two conductors and a shield/ground. The two conductors carry the two phases of the electric signal. These are protected from inductive interference by a conductive shield that also serves as system ground. If the two phases (+/−) are combined, they cancel each other out. They are exact opposites (180 degrees out of phase).

Unbalanced cables are more susceptible to inductive interference. The unbalanced transmission cable has one conductor and a shield. The single conductor is carrying the positive (+) side of the electronic circuit, and the shielding is sharing the negative (−) side with the ground. The RCA, TS, and the BNC connectors are perfectly appropriate for this transmission format. Because the negative (−) side of the circuit is traveling on the shielding, it is effectively unprotected from inductive interferences, and therefore best used for short transmission distances.

Shielding is the only effective way to protect the conductors and the device circuitry from radio frequency (RF) interference. If you hear "ghost" signals of radio broadcasts in your program, it means that the integrity of the transmission is improperly shielded. It could be faulty connections, shielding, or cables.

2.12 Grounding

Grounding is the electrical connection to the ground—literally, to the earth. The electrical power of your studio or home has three conductors. This is a balanced power system where alternating current is transmitted along two conductors and the system is referenced by a third wire to the earth. In electrical circuits, this same reference to the earth (ground) must still be maintained. If you hear a constant "hum" in your audio, there is a possibility that there are multiple paths to ground (a ground loop). The ideal system has one (and only one) path to ground for all electronic components. It is highly advised that the AC power for all electronic devices be plugged into a common power strip or from a common grounding point (star grounding). If you still find multiple paths to ground, there are AC power adapters that have two conductors and a floating ground (ground lift). Alternatively, you can isolate the ground for each device until you have removed the offensive ground loop. This is not recommended unless you are an expert because it is potentially dangerous. If you are having problems with grounding, seek professional help from an electrician first.

All of these issues help describe why the professional world takes great care in designing an electrical/audio/video transmission system that provides for the integrity of signal conductors, phase coherency, shielding, and grounding for each and every component of a complex system.

Professional studios sometimes use unbalanced cable with their gear when they have control over possible interference. Using unbalanced cabling, while a risk, reduces the need for extra electronics in equipment needed to reverse the polarity of the cables. This results in a higher quality signal. Live sound, however, depends on balanced cables to get as clean a sound as possible. There is a lot less control over interference in large venues, considering the nature of the sound systems in that kind of environment. Microphone cables deal with such low-level electrical signals that they require balanced cables because they are more susceptible to interference. Alternatively, speaker cables deal with such high levels that they rarely need to be balanced. Consumer gear, which you will inevitably use, uses unbalanced cables, and care should be taken to prevent interference.

You can do this by:

- Keeping sensitive cables away from other cables and power sources
- Using shielded cables (with insulation)
- Using digital connections, which are more immune to interference noise, rather than analog inputs/outputs
- Running cables perpendicular to possible interferences, as this will help prevent serious problems
- Using optical cables for digital signals when possible

2.13 MIDI connections

In a book about sampling, it is a good idea to review MIDI connections. MIDI stands for *music instrument digital interface*. MIDI is a control protocol that is used much like a roll of paper in a player piano. All modern electronic keyboards use MIDI to indicate such things as when a key is pressed, how hard it is pressed, and when it released. MIDI can also capture continuous controller information, such as pitch bend, modulation, and after-touch. Each of these is discussed in other sections of this book, but what you should know about making MIDI connections is listed here.

The MIDI format has its own unique cable and connector. What is currently considered a MIDI cable originated in the 1960s as an audio cable but is now rarely used for anything but MIDI. The connector is circular and has five pins (Figure 2.15). Ironically, the cable only uses three of the five pins because room was left for expansion. The format worked so well that it went relatively unchanged for several decades. Instead of upgrading the standard MIDI cable, newer formats have been adapted that allow additional functionality. Even though there are better and faster ways to do the same thing, five-pin MIDI cables are still widely used.

A number of modern MIDI controllers use USB cables to connect to computers. If you need to hook up a non-USB controller to your computer, then a good solution is to use a

Figure 2.15 MIDI Cable.

USB MIDI interface. This allows you to hook up a five-pin MIDI cable to a computer. There are also several audio interfaces that have MIDI I/O built in and a number of MIDI-only interfaces that have ten or more MIDI in/outs. There are other controllers that connect wirelessly to computers, eliminating the need for any cables.

When using standard MIDI cables, it is possible to connect several pieces of MIDI equipment together in a chain. Most samplers and synthesizers that use MIDI will have three MIDI ports: in, out, and thru. Any MIDI signal that enters the input is automatically sent straight to the thru port. This can be sent to another piece of equipment so that a single controller can communicate with several pieces of gear at once. There is a slight delay and degradation of the MIDI signal associated with connecting MIDI equipment in such a series, but this is only an issue when hooking more than a few pieces of gear together. It is recommended that three or fewer devices be hooked up in this manner.

2.14 General equipment tips and tricks

As you may have noticed while reading this chapter, we have not provided a long list of equipment specifications and super-detailed explanations on how things work. The primary goal has been to help you apply the vast amount of information already available in the context of sampling. So, instead of filling up a lot of pages with information you probably already have, this section has been quite limited in scope. On this book's website there is more information about specific equipment. This final section is the most general section of the entire chapter, but it contains several pieces of advice that should help you during the sampling process.

Web forums

2.14.1 Forums

When it comes to troubleshooting, it can be quite difficult to find solutions. Anyone who has had the pleasure of dealing with online or phone-based technical support can usually tell at least one horror story about spending hours on the phone and in the end getting no help at all or making things worse. Quite often, product manuals are not much better. In nearly every case with audio technology, there are many other people going through the same challenges. On the website for this text, you will find links to support forums for nearly every topic related to sampling and sampling equipment. These include discussions ranging from computers and microphones to cables and field recording. It isn't practical or possible to create a document that explains every single problem with possible solutions, so instead you can go to a single website and access links to quite a few different forum locations. The Internet is a great tool for bringing people from all over the world into a single conversation about technical problems.

2.14.2 Patience in buying new gear

When buying software and audio equipment, do not go beyond your means to get the latest and greatest stuff. An article written by Rich Sanders for the industry magazine *Recording* (November 2003) entitled "How Not to Make Money in the Recording Business!" describes how some people buy equipment based on the expectation that they will find a place to use it. Instead, if you plan on creating samples for money, wait until there is a specific need and then buy the equipment that fills that need. Not only would it be a justified purchase, but hopefully it would be paid for by the project. Don't misunderstand this suggestion: If you are creating samples for fun, then get whatever you can afford. However, if creating music is part of your livelihood, be careful. In an industry where new technology is always being created and where new products are never ending, it can be very difficult to be satisfied with last year's model. There are several prominent engineers who consistently use an older set of tools long after everyone else has switched to the newest thing. Their tools do not limit them because they know how to use them. They do so while others are constantly learning new technology. When their tools begin to prevent them from accomplishing what they need to accomplish, then they replace them. Think of the money saved by not switching systems every year. Home studios can really benefit from such a paradigm.

2.14.3 Find a quality reference

In the vein of the previous paragraph, find a single tool from each category of tools that you really like. Learn everything about the tool, including how to use it and how it sounds. It might not be the best of the best, but it will be something on which you rely for results. A good microphone is an example. Find one that you know how to use to get the results you desire. A large diaphragm condenser that is relatively transparent would be a good choice. You need a quality reference for comparison as you add new tools to your collection.

Pay attention to the details. Know each piece of equipment that you use, all the way down to its detailed specs. If you own some equipment that you don't understand, then look up the information online or in a book. In many cases, you can call someone at the company and ask questions. Keeping a notebook with information that you come across will save you time and energy in the long run.

SOURCE PREPARATION 3

With so many options available, it is wise to establish the primary goal of your sampled instrument creation. This involves choosing and preparing your source material before you actually begin recording samples.

The good news is that you can put as much or as little time and energy into this project as you want. You can make a simple instrument fairly quickly (under an hour), but a complex instrument can take a lot of time and organization. From the project inception through the recording and mapping, a complex project could take several days. This chapter will help you focus your ideas and prepare you to make decisions about what you would like to create. This chapter will also help you lay the foundation of a successful project by showing you what it will take to prepare the sound source that you decide to use.

3.1 Choosing your source

3.1.1 Traditional instruments

It is possible to create many different types of sampled instruments. If you would like to use an existing instrument as your source, then many of the things you will do to create it are laid out before you. Using a violin as an example:

- There are a fixed number of notes that a violin can play.
- If you play a violin in a traditional manner, then it will use a standard tuning system.
- The strings of a violin can be played with a bow or with the violinist's fingers.
- Most importantly, a violin has a sound that is recognizable. If you attempt to create a sampled violin, then it should sound like a violin (Figure 3.1).

Figure 3.1 Violinist.

3.1.2 Brand new instruments

Another option is to create a brand new instrument. Take the same violin, but this time use a piece of metal and scrape it along the strings. There is a good chance that most people wouldn't be able to tell you what instrument is being played without looking. You could also create a new percussion instrument by using non-traditional items. Plastic containers and cupboard doors could be used as drum sounds. It quickly becomes clear that when it comes to creating new instruments, there are very few limitations.

3.1.3 Existing vs. new

Decide from the beginning whether or not you want your creation to be a realistic representation of an existing instrument or something fresh that is being made for the first time.

Some examples of possible instruments are:

- Traditional instruments
- Orchestral (stringed, brass, woodwind, percussion)
- Pop/rock (guitar, bass, drums)

- Electronic (synthesizer, electric piano)
- Non-traditional
- Machinery
- Things with vibrating parts
- Kids' toys (any toy that makes sound)
- Household items (too many to list)
- Anything in the world that makes sound

The traditional category is often much easier to work with conceptually. You can find recordings of what those instruments are supposed to sound like. You can also find charts of their ranges and how to tune them. The non-traditional category is more open ended. For your first instrument, you should choose something manageable. Once you decide on a source, the preparation stage will have just begun.

3.1.4 Who should be the performer?

When deciding on what to use as a source, you should consider the two performance options. First are sources that you can play yourself, and then there are sources that someone else will play. As you may realize, there are advantages to doing things yourself. You can spend all the time you want without worrying about someone else's schedule. You can work whenever and for however long you want. You don't have to worry about performance nerves, because you can play the instrument by yourself with no audience. The list goes on and on. However, most people specialize in certain instruments. I highly doubt that you are a world-class pianist and a world-class trumpet player and a world-class clarinet player, and . . . well, you get the idea. If you choose as your source an instrument that you do not play very well, you should consider recording someone else and not doing it yourself.

Can you have your non-professional musician friends play them? Because you will be recording only a note at a time, which isn't too difficult, the simple answer is yes. However, if you want to make the next best-selling sampled piano and you plan on selling it worldwide, you probably need to record a very nice piano, in a nice room, with a skilled piano player. There are so many examples of samples in between excellent and terrible that it really comes down to what you want to do and what you have available to work with.

When it comes to creating an instrument for the first time, it is highly likely that you will perform it yourself. There is no rule either way, but you are the one who will be taking the time to figure out how you want it to sound, and you will be the one who has the idea in your head of how you think it should or could sound. Additionally, you may be new to the sampling process, and performing the sound source by yourself means that someone else won't have to sit around while you experiment. In either case, a performer will be needed, and it is up to you to decide the best way to meet your end goal.

3.1.5 Non-acoustic sources

One type of sound source comes from anything not recorded by a microphone. This includes sources recorded by others, or sources that simply need to be converted into a digital format. For example, you could use prerecorded sound files to create your instrument, which you can find online or on CD-ROMs that come with magazines. If that is your source, you need to make sure you aren't breaking copyright rules in the use of these materials. This is especially important if you plan to use your newly created instruments in projects that will be released commercially. Even if you are using your instrument strictly for personal projects, it is still important to avoid breaking the law.

The Internet is a great resource, and there are sites that let people post recordings for everyone to use. Some search engines allow you to search for audio files, and you might be surprised at what you can find. I recommend staying clear of posted Internet files unless it is obvious that the files are free to use.

Sampling: Pay $ Now Or Pay $$$$ Later

By Storm Gloor

Use of samples can greatly enhance your recording and possibly provide the "hook" that makes a track a hit. But don't forget that almost every sample has a rights holder/copyright owner. You're *far* better off negotiating for the clearance and terms of use of the previously recorded work—no matter how much of it you use— *before* you finish your recording. Taking your chances, assuming the use is negligible, or waiting until the recording is released can be very costly. Both the producer and the artist may suffer damage to their pocketbooks. Once a recording is released with un-authorized samples, copyright owners have the upper hand in any negotiations to correct the situation; they can also seek preventive legal means to recall the product from any company that distributes or retails the recording—a potentially ugly mess for all involved that can hurt the artist's and the studio's reputations. In fact, if the record is a hit, you can bet that the owner will be seeking a much larger share (if not all) of the profits, which may not have been the case had authorization been cleared before the recording was even mastered.

Negotiations for use of the sample can result in some fees or costs—for example, a buy-out of rights fee, a certain percentage of the income from the track, a transfer of a portion of the copyright ownership of the new recording—or some other arrangement, which may combine some of those options. The nature and duration of the sample, and how it will be used, will be factors. And, again, requesting approval *prior* to its use (the earlier the better) will work in your favor.

It is not safe to assume that because the sample is very short or hardly recognizable, there is no need for clearing the use of it. There are no hard-and-fast rules as to the precise length that is a "fair use" of a sample. And you can bet that publishers and copyright owners always have their ears open to identify any uses of their material. When it comes to sampling, it is much safer, and likely much less expensive, to ask permission early rather than later.

Sources:

Brabec, Jeffrey, and Todd Brabec (2006). *Music, Money, and Success: The Insider's Guide to Making Money in the Music Industry*. 5th ed. Schirmer Trade Books, New York.

Halloran, Mark (2007). *The Musician's Business and Legal Guide*. 4th ed. Prentice Hall, New Jersey.

3.1.6 *Straight in*

Another way to capture sounds without microphones is to attach a synthesizer (or guitar, bass, etc.) straight to your sampler and record the output.

If you are using a digital keyboard (Figure 3.2) and capturing an exact copy of the sound, then it is no different from stealing the sounds. If you own the original sample, then it is okay if your copy is strictly for personal use; it would be a problem only if you were pretending you made the sample yourself or if you gave it away to someone else. If you are using a synthesizer and recording the output, then you are generally safe because the synthesizer is not using a prerecorded sample that you are duplicating. The synthesizer is actually creating a brand new sound each time you play it, and you are allowed to record it for use in a new sampled instrument. A good rule of thumb is to refrain from using other people's work verbatim unless you have express permission to do so.

There are programs (Samplerobot and Autosampler) that can create a sampled instrument from any MIDI-controllable instrument. These programs send out every possible MIDI

Figure 3.2 Keyboards.

note to the destination instrument and then record the sounds and map everything into a new sampled instrument. This is very efficient but can violate copyright laws. The primary purpose of this software is to create a software version of hardware instruments that you already own and for which you want a portable, software version. It is not designed to enable you to steal sounds from instruments you do not own.

3.1.7 Batteries of non-pitched instruments

If you decide to create a non-pitched instrument, you should be aware that you can create a *battery* of sounds. This is an instrument that has a different sound assigned to each note of your sampler. In fact, the process is flexible enough that you can make any key assignments you want with pitched and non-pitched instruments. An example of a battery that you could create is an instrument that includes sounds from hitting various windows. Go through your house testing windows for a possible instrument. A window acts very much like a drum head in its vibration tendencies. Choose a beater to tap the window, and compare different windows to one another. You can then record each window and map them to different notes. Middle C could be the living room window. B could be the bathroom window. Bb could be the sliding glass door. You get the picture: this can be a very flexible process. If you decide to use windows as a sampled instrument, be careful when striking the glass, because it can be dangerous if the glass is broken.

In the same vein as the battery concept, you can also create performance patches or splits. These are commonly used when performing live. You can make half of the instrument a piano and the other half a bass. When creating your own sampled instrument, you can put any combination of sources together to form one entity. You should probably base your decisions on the practicality of what your goal is. If splitting your instrument into four parts with a very strange combination of instruments is what you are looking for, then do it.

3.1.8 *Focus is on musical instruments*

In this text, we are focusing on the creation of sampled instruments that are designed to play music. The techniques we cover, though, could also apply to any end goal that you want. Samplers can be used to trigger SFX or full music tracks. A key can be used to trigger laughter for a television sitcom, or to play the latest pop tune on a radio station. Samplers are very flexible and their uses are very broad. Do not hesitate to push your sampler to the limit as you explore what is possible.

3.2 Preparing your source

After you have decided which sound source to use, you should prepare the source for recording. The physical preparation of the source is important because once it is recorded, it is hard to fix a lot of the problems that might otherwise have been avoided. Start with a source that already sounds like it should, and you will have a much better chance at creating a sampled instrument you are happy with.

Even with the best source instrument, it is possible to have poor results if it is not prepared properly. Never assume that if the instrument belongs to a professional musician that it will automatically work in a sampling environment. You have to take some steps to ensure success. The more detailed the preparation, the better and more efficient the process will be.

- Make sure all moving parts are maintained. Lubricate moving parts that are squeaky. Replace pads where needed to avoid metal clanking and/or excess breath leakage if applicable. Oil valves and slides. Check all moving parts for extraneous noise. Of course, if you would like any of these additional sounds to be a part of your sampled instrument, then do not do anything to change them.
- If it is a stringed instrument, use strings that are not worn out.
- For drum kits, it is good to use fresh drums heads.
- Make sure the instrument is tuned properly and continuously during the recording process.
- Some sources might need to be dampened if they are too resonant. If you are using a hollow item to create a drum sound, it might need some additional mass placed inside to help dampen unwanted sound.
- Make sure all necessary accessories are available. This includes drumsticks, mallets, batteries, tuners, and the like. You can never be too prepared.

With brand new instrument creations (for example, using as the source household items that are not usually used as musical instruments), you will have to exercise your best judgment on the physical preparation of the source. There are no rules governing how such an instrument should sound, and any "extra" noise may be a part of the instrument. There is some wiggle room here for creative license.

3.2.1 The range of your virtual instrument

When creating a virtual instrument, you are in complete control of every single detail. The range of the instrument is an example of such a detail. Imagine a virtual violin that can play higher and lower than an actual violin. It might not be the best option, but it is possible. Alternatively, you can create an exact replica of the instrument. In some cases, you might want to use a partial range. You would do this to save time in the creation phase, especially if you know you will never use a particular part of the range.

3.2.2 What is the range of a kitchen cupboard?

Creating an instrument that is non-traditional demands that you look at the instrument's range in a different way. Let's say you want to make an instrument that uses cupboards as the source. What is the range of a cupboard? That's your call. You're the creator, so it only makes sense that you get to define the range. An example range in this case could include different styles of shutting the cupboard door. You could have the slam, the gentle close, the close and then hold, or the close and then a bounce. It might also include different opening techniques that could encompass a slam into a neighboring cupboard. The point is that there is a lot of room for creativity, and you can do whatever you would like to do. Don't be afraid to try new things, and never assume that you have to create a traditional instrument.

3.2.3 Tuning issues

The equal-tempered tuning system is the most common tuning system used with sampled instruments. The equal-tempered system takes each octave of the instrument and divides it into twelve equal steps. Is this the best way to tune an instrument? That depends. When the pitches are divided this way, there are a few tuning irregularities such as the 3rd and 5th scale degrees not being perfectly in tune with some of the chords. This means that pianos (which use the tempered system) are all slightly out of tune. The equal-tempered tuning system allows a piano to be played in every key. This is a system built on compromise. Another system of tuning, called *intonation*, is not split evenly over each octave. This system takes into consideration that not every interval is the

Tuning system charts and links

same. Chords are more in tune, but with instruments that are not easily tunable, it is hard to play in more than one key. The equal-tempered system, on the other hand, is designed to allow instruments to play in all possible keys without constantly retuning every note. This system allows the tuning to be fairly close despite never being exactly in tune.

3.2.4 Instruments can be more or less in tune

With your new musical instrument, it is possible to tune it any way you want to, and it can actually be more in tune than an original source. An interesting fact is that most source instruments can play perfectly in tune. A violin has no frets, and so violinists can simply move their finger until the note is in tune. Trumpet players can raise or lower a pitch by adjusting the buzzing of their lips. When instruments that can be adjusted by the player are sampled, they are typically brought into the equal-tempered tuning system and are no longer played perfectly in tune. This is one reason that many sampled instruments do not sound as realistic as their physical counterparts.

3.2.5 Equal-tempered tuning is easiest

You can choose to go either way on this, but most samplers tend to lean toward the equal-tempered tuning system because it is easier. Creating an instrument that is in tune for a single scale is a great idea, but imagine having to create the same instrument for all twelve scales. You also have to think about all of the other modal variations. It could become a very cumbersome project, and not very practical in situations where the instrument will need to play in multiple keys.

Tuning traditional instruments is easier because there are set standards. When you decide to work creatively, things can be a little trickier. An example of this is a toy xylophone that could be made into a sampled instrument (Figure 3.3).

During the recording process, it would be very obvious that the pitches are not in tune with any known system. Some notes might be extremely out of tune, as this is typical for a

Figure 3.3 Xylophone.

toy xylophone. Is it possible to tune a xylophone like that? It's not very easy. With the myriad of tools available, it is possible to do a lot with intonation changes. In this case, a pitch-shifting plug-in could be used to adjust the pitch of the notes that are out of tune. After completion, you could have a very good sounding children's xylophone that can be used as a creative tool when creating music.

When getting an instrument ready, you need a tuner to help you. Many sources can be tuned, and a tuner can assist in this process. You can use a hardware tuner or a software tuner. Many digital audio workstations have tuners built in.

3.3 Understanding your source

There are many books that offer information about the fundamentals of sound. This section is not designed to replace the extensive literature on the subject but is meant to help in the context of creating samples. The goal is to help you understand how you can work with different sound sources in a variety of situations.

3.3.1 Change in pressure

Sound is a change in pressure. When you clap your hands, a wave of energy is created and a chain reaction carries this energy outward from your hands. The air around your hands moves. As it does, air particles bump into other air particles and create movement. As the air particles collide, an area of dense pressure is created which then forces expansion.

This chain of events travels through the air, but sound can also travel through many other materials. It can travel through walls, floors, doors, water, rocks, and other materials. It travels better through some materials than others. Sounds from a source typically travel through the floor faster than through the air. This creates an interesting effect when you feel the sound in your feet slightly before you hear it with your ears.

3.3.2 The ear

When a sound enters your ear, the eardrum helps translate the pressure differences to a signal that your brain understands as sound. The pressure changes in the air work in a cyclic system. There is a section of compression and then one of expansion (Figure 3.4).

Figure 3.4 Waveform.

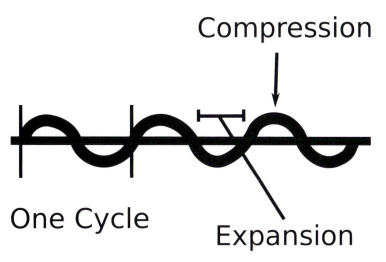

These cycles are measured in hertz (Hz), which means cycles per second: 1 Hz describes an area of pressure expansion and compression that takes 1 second to complete. In comparison to the cycles in the typical range of pitch recognition, 1 Hz is very long.

The human pitch recognition range is from about 30 Hz to 13 KHz. We can hear beyond these limits, but there is no pitch recognition. There are many things happening around our ears that create changes in pressure, but only those that are repetitious are translated as pitches. If your ear receives a bunch of non-related pressure changes in the range of hearing, it is most likely a form of noise or non-musical sound. If the ear receives a string of pressure changes that focus around a specific frequency, then it will be interpreted as a pitch.

3.3.3 *The speed of sound*

Even though the pressure changes take place at different rates, the overall speed of the wave movement is limited to the medium in which it travels. The speed of sound in air is approximately 343 meters per second at normal room temperature. This is quite a telling piece of information, because now that you know the speed of sound in the air, you know the length of any sound wave.

Let's do some math. If you are listening to a pitch that is 343 Hz, then there 343 pressure cycles each second. The sound itself can travel 343 meters during that 1 second, which means that each individual cycle is 1 meter long. To find the length of the wave, take the speed of sound and divide by the frequency: $343 \div 100 = 3.43$. This means that a 100 Hz pitch is 3.43 meters long. Another way to look at it is that it takes at least 3.43 meters before one cycle of the sound has fully developed.

A minimum expectation for the frequency range of professional gear is 20 to 20,000 cycles. A 20 Hz tone is how long? Calculate $343 \div 20 = 17.15$ meters. That is fairly long. On the other end of the spectrum, we have 20,000 cycles: 343 ÷ 20000 = .01715 meters, which is very short. This is one reason why people say, in reference to acoustics, that low frequencies are one of the bigger problems. At 17.15 meters, they are literally a "big" problem.

3.3.4 Applied knowledge

Why does all of this matter? This text tries to focus on applied knowledge and, while this discussion of pressure changes in the air might seem slightly academic, it is really quite valuable. The standard tuning reference uses 440 Hz for the note A. This is the first A above middle C (Figure 3.5). This automatically determines everything else. An octave above that A would be twice the frequency, or 880 Hz. An octave below equals 220 Hz. The key to understanding the equal-tempered system is to understand ratios. An octave can be expressed as a 2:1 ratio. The notes in between are split into 12 equal parts. The math behind this gets slightly more complex, and the frequency ratios between pitches are quantified as the 12th root of 2. This means that when a frequency ratio between adjacent notes is multiplied by itself 12 times, it equals 2. The important thing is to realize that equal-tempered tuning allows you to play in all keys with minimal intonation problems. For more information on this see Harry Partch's *Genesis of a Music* (2nd edition, Da Capo Press, 1974).

When using a source like a window, you can deduce, without ever touching the glass, how its pitch will relate to that of another window. The bigger window will have a wider vibration and create a longer pressure disturbance. With sources that vibrate like strings, understanding acoustics will inform you that tighter strings translate into shorter vibrations. Shorter vibrations create a shorter pressure disturbance and the pitch will be higher. If you cut the string exactly in half, it will be exactly an octave higher. It might seem that the higher pitches should be traveling faster, but they are really just shorter and so more fit in a given time period. All sound in the air travels at the same relative speed.

Figure 3.5 Notes and Frequencies.

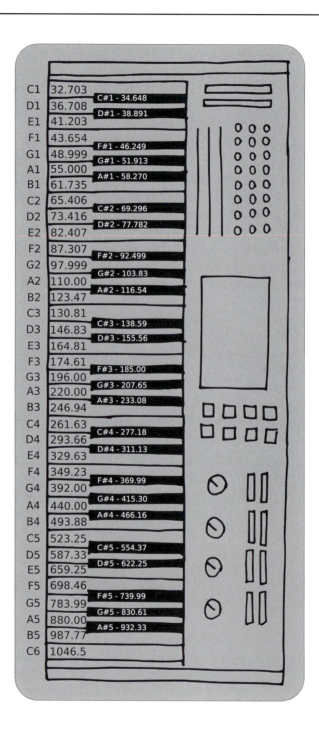

Figure 3.6 Typical
Instrument Ranges.

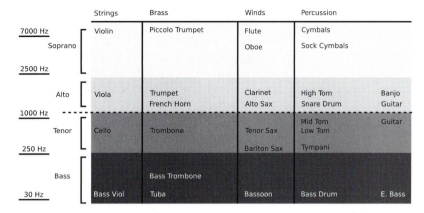

Ranges of Musical Families

When you work with a digital audio editor, all of the same principles apply. The graphical display is designed to graphically represent the pressure changes in the air. The visual waveform shows cycles, compression, and expansion. You will see these graphical representations throughout later chapters.

Once you understand the basic principle of how sound works, you should be able to work more efficiently at your source preparation. Many obstacles to working with pitched instruments can be figured out intuitively based upon the principles listed here.

3.4 Putting it all together

We have covered source choice, source preparation, and source comprehension. As you work with your source, you will often run into obstacles that work against you in the creation phase. Extraneous noise, being out of tune, and poor sound quality are just a few of the issues you might face. In these cases, you will need to be creative and resourceful when looking for solutions. Once you have picked your source and prepared it, you can begin the recording phase of the process. The recording phase does not end the preparation phase, because you will be constantly working with the source to ensure that it sounds as good as possible throughout the entire process.

RECORDING THE FIRST SAMPLE   4

4.1   Ear training

Before we discuss the actual recording process, we are going to spend some time discussing ear training. Ear training is the process of developing your ear through consistent practice. Ear training is a skill that will help improve the quality of your finished instrument more than any other skill. The success of ear training, like any type of training, depends on having clear objectives and then engaging in consistent practice. The primary objective of this section is to provide you with a practical approach to developing your ear to enhance the sample creation process. It is recommended that you make a study of physical acoustics as well, and a bibliography has been prepared to help you find good sources.

A good ear is the single most important factor for understanding your acoustic space. Training your ear will give you the necessary tools to create the samples you have envisioned. It should be noted that a good ear is not a critical part of creating samples. A sample can be created from beginning to end without ears by using only visual cues. The results might not be very good, but it is possible.

A trained ear empowers the decision-making process. Do you want a hyper-realistic sample of a grand piano? How do you know when you have recorded enough individual samples to re-create the whole piano? Are you recording enough of the right information? Are there certain parts of the sound source that are critical to the sound but are not being captured by the microphone? It is amazing how a sampled piano

are good at looking at lips to help them understand what is being said. If you listen to a recording of the same thing, it is often impossible to understand anything because there are no visual cues. Once you make the comparison between what you hear and what was recorded, you should have a pretty good idea of your transparent listening ability. The more you do this, the better you will get at it.

4.1.3 *Focused listening*

Focused listening is the process of listening to specific sounds while ignoring others. This is how humans normally listen, and it is instinctual. However, actively choosing what to focus on requires self-discipline and additional training. There are many situations in which your ears automatically focus, but when you have sound that you want to analyze in a specific way, you may have to force your listening to focus on that specific aspect. Usually, focusing is not the problem; it's staying focused over a period time that takes practice. Endurance listening in a focused manner is not easy. Focused listening needs to be balanced with transparent listening. If you listen only in a focused way, some things will be missed. If you are recording in your living room, you might miss the sound of the refrigerator from the kitchen. The refrigerator is a great example because when a sound is constantly around, your mind has the ability to tune it out. If it isn't a threat and it's not important to you, then it gets ignored. This ability is good for your sanity, but it is bad for recording. While recording, you might miss the sound because your focus is on the source, but later when you listen back to the recording, the noise will be more obvious because you are focused on the recording in a different way. Keep a balance of transparent and focused listening.

Why is focused listening important?

Focused listening helps break through the effects of casual listening. If you have a source that you have heard many times, it is typical for your mind to get used to it. Any flaws or characteristics that exist as a part of the sound become normal to you, even though they might be considered "wrong" or "bad" by accepted standards. Focused listening helps you analyze the different parts of the sound. The ability to hear the sound with a high level of scrutiny allows you to truly hear the sound.

How is focused listening developed?

Here are a few ways to develop focused listening:

1. Listen to a song and pick one part to listen to in a critical way. Take a bass part and listen to it while ignoring the rest of the song. You may start with a vocal part, but keep in mind that the vocal lines are designed to be the main focus and they are easy to keep in your focus. If you are having a hard time, work your way from easier parts to more obscure parts. Keep track of what they do and what they sound like.

2. Sit in a place where there are sounds all around you. Start with a transparent listening exercise and listen to all of the sounds equally. Start to focus in on the sounds, and draw a map of all the sounds and where they are located. Once you have a map drawn, take time to focus on each of them individually. Break them down into different categories. Some may be high pitched and others low pitched. Some might be noise and others might be musical. Some may be machine-like and others more organic.

3. To develop endurance in focused listening, pick different sounds that you hear and use a watch to time how long you can stay focused on them. If you do this exercise consistently, you will notice your focus growing longer. It also helps develop your ability to focus more intensely over shorter periods of time.

4.1.4 Listening memory

The ability to remember what something sounds like is the next skill in the set. Listening memory is sometimes under-developed among people who record sound, because they rely heavily on the recordings instead of using their memory.

The recordings can act as unneeded supports that prevent listening memory from being fully developed.

Why should listening memory be developed?

A sound captured by a microphone is not necessarily an accurate representation of the original sound. There are so many variables involved that it is not wise to simply trust that the recording sounds true to the original. Listening memory is a key element in guaranteeing accuracy in the recording process. If you hear the sound and then hear the recording, you should be able to tell if they sound the same. A well-developed listening memory helps you remember how something sounds, even after a period of time has passed.

How is listening memory developed?

Take a moment to try the following exercises for developing listening memory:

1. This first exercise is more for developing general auditory memory. Listen to a song and try to remember everything about it that you can. Once it is over, make a list of what you can remember. This includes as much information about the type of instruments that were used, the type of vocals, song structure (how many verses, choruses, bridges), and anything else that is unique about the song. Once you make the list, then relisten to the song and add things you missed to the list. This will help lay a good auditory memory foundation.
2. This next exercise takes a little more work to accomplish. You will need a recording system, as you will be listening to an original source and comparing it to a recording. Pick a sound source you would like to use as an example. Make a recording of it, and then listen to it and then the source a number of times until you are able to identify the similarities and differences. Make a list of what you notice. What are the similarities? What are the differences?
3. Once you have compared the original to the source, go back and record the source as many times as necessary until the source and the recording are as similar as possible.

This requires some experimentation with microphones and positioning.

4. Attend performances of an instrument (or ensemble) in a natural environment. Listen analytically, and listen from a variety of locations. Give your ear the benefit of experience. Invest in your ear training.

4.1.5 Projected listening

Projected listening is the ability to predict what something will sound like, either in a particular place or on a recording. While this is a skill that can be developed, it is hard to quantify the process of growth. Projected listening is related to listening memory, but it is not exactly the same. If you try to imagine what people you know will sound like when they open their mouth and speak, then there is a good chance you are using your memory and not projecting a new sound. The memory aspect is active here because you are remembering what the voices sound like, based on previous listening and not on something new. When you try to imagine how a person's voice will sound in a large cathedral, you are taking several pieces of information from your memory and placing them together to form something new. This combination of listening memories to form something new is the key to projected listening.

Another example is when you are creating a brand new instrument. If you have an idea for a specific sound, then you will try to find a way to realize the sound by creating an instrument. Projected listening is a skill that will come in very handy during the creation process and as you plan and construct new instruments. Instead of creating a new instrument based on trial and error alone, you can make educated guesses on how the new instrument will sound.

Why should projected listening be developed?

One of the primary practical uses of projected listening is to help you create a recording that sounds like the original sound source. If a recording doesn't sound like the original, you need to find a way to create consistency between the two. You can experiment with a number of variables until it sounds right, or you could project, in your mind, a solution that would solve the problem without spending as much extra time. The ability to find a solution without too much trial and error can be a huge time-saver.

How is projected listening developed?

The following exercises can help you develop projected listening skills:

1. Listen to established instruments in their original environment. Spend time attending concerts and performances, and listen analytically.
2. Pick a sound source that is portable and take it to a number of different acoustic spaces. Before listening to the source in each of the places, make guesses about what you will hear. While making your guess, look around and analyze the space as best as you can. Once you listen to the source in the space, compare what you thought would happen with what actually happened. If you were wrong, try to figure out why.
3. Gather a group of objects together and create a musical instrument. Before you play it for the first time, guess what it will sound like. Alternatively, try to make two physically dissimilar instruments sound the same.

4.1.6 Summary

All four of these skills are interrelated. Transparent listening is the starting place for focused listening. Listening memory can only take place after transparent listening and focused listening have taken place. Projected listening is only effective if your other skills are fully developed. While it is important to train your ear with these skills in mind, you should remember that the skills are a means to an end, not the end itself. Training your ear will help you create samples that sound like you want them to.

How does ear training help with acoustics? Your ear will inform you of the appropriateness or inappropriateness of the acoustics surrounding your sound source. If the source can be moved, then your ear will help you find the appropriate place. If the source is fixed in place, then your ear will help you change the acoustics around the source until they are appropriate.

Acoustic resources

4.2 Basic acoustic issues

As you train your ear, you will begin to notice how acoustics affect the sound around you. You will notice that some rooms sound "better" than others and that specific source placement changes within a room can make a significant difference. Sometimes it is possible to eliminate poor room acoustics through microphone techniques and through treatments, but in other cases there are no remedies. You will also learn that sound in a free field, such as someplace outdoors, has very inconsequential reflections and is void of "room sound," but there are often too many other sounds that create interference to be ideal for recording.

Acoustic issues and problems often stem from *reflections*. In rooms that are cube shaped, the reflections between the walls, ceiling, and floor can cause harmonic coloration and an uneven frequency response. Both of these can adversely affect the sound of your source. The size of the room determines the frequencies that are affected, and you can easily figure out those frequencies using the wavelength formulas from Chapter 3. The distance between two surfaces reinforces a specific frequency. A room that has multiple surfaces that are the same distance apart boosts the frequencies and harmonics that correspond to specific distances. Parallel walls with reflective surfaces act as a reflector for higher frequencies, creating a slap echo between the surfaces. This results in discoloration and a general smearing of the sound in the air.

Solutions to acoustic problems include picking rooms with desirable measurements and/or using acoustic treatments such as reflectors, diffusers, and absorbers.

The ear training methods described in Section 4.1 barely scratch the surface of available research and known acoustic principles. They are designed to help you train your ear so that you can make decisions based on results and not specific acoustical design. Please see the following sources for more information and guidance.

- Cavanaugh, William J., and Joseph A. Wilkes (1999). *Architectural Acoustics: Principles and Practice.* Wiley, New York.
- Chappell, Jon (2006). *Home Recording Studio Basics.* Cherry Lane Music Co., New York.
- Everest, F. Alton (2001). *Master Handbook of Acoustics.* 4th ed. McGraw-Hill, New York.
- Gallagher, Mitch (2006). *Acoustic Design for the Home Studio.* Thompson Course Technology, Boston.
- Howard, David, and Jamie Angus (2006). *Acoustics and Psychoacoustics.* 3rd ed. Focal Press, London.
- White, Paul (2003). *Basic Home Studio Design.* Sanctuary, London.

Audio examples

4.3   Microphone choice and placement

Microphones are one of the key elements in a recording situation. Effective use of a microphone is more important than the type or price involved. Microphone choice and placement should be based on your goals for each individual situation. Consider the following two scenarios in the decision-making process.

- Capturing the sound source and the ambience around the source
- Capturing the sound source while eliminating the ambience around the source

4.3.1   Capturing source and ambience

Microphones are described here in terms of their polar response (polar pattern). These responses always vary with frequency. Expensive professional microphones will have a printout of the polar response at several frequencies. The lower the measured frequency, the less directional the polar response.

All of the microphone types can be used to capture the source and ambience around the source if used in specific ways. Omnidirectional microphones are the obvious choice for recording everything around a source, because they accept sound from all directions.

If an omnidirectional microphone (Figure 4.1) is placed in front of a source, then it picks up the sound from the source and the ambient sounds from around the source. If the microphone is very close to the source, then the source will be louder than the surrounding ambience. The farther you move the microphone from the source, the louder the ambience will seem in proportion to the direct sound from the source. If the distance is continuously increased, there will be a point where the ambience and direct sound will be equal. Eventually, the direct sound will decrease beyond usable levels.

A bidirectional microphone (Figure 4.2) may also be used. This type of microphone accepts sounds from two opposite directions and everything else is rejected. If a bidirectional microphone is placed in front of a sound source, the microphone will reject any sound coming from the side directions. This can be a great tool when there are distracting sounds

Figure 4.1 Omnidirectional Pattern.

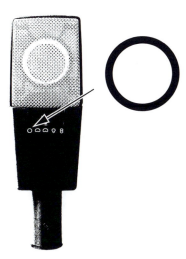

Figure 4.2 Bidirectional Pattern.

coming from a side direction but you still want to capture the ambient sounds around the source.

Unidirectional microphones (Figure 4.3) pick up sound from one direction and reject sound from other directions. This type may not seem like a good choice for ambient sound capture, but it can work well in certain situations. If you cannot get near a sound source, then using one of the other patterned microphones will pick up too much ambient sound in relation to the direct sound. Unidirectional microphones allow you to be farther away because they reject most of the ambient sound and focus in on the sound source. If you are far away, however, the unidirectional microphone allows you to pick up the ambient sound immediately surrounding the source, just as if you were using an omnidirectional microphone much closer to the source. You need to test this carefully, however, as unidirectional microphones are not consistently the first choice to capture ambient sound.

4.3.2 Microphone placements for capturing source and ambience

A single microphone may be used to record the sound source. The amount of ambience you would like will determine the

Figure 4.3 Unidirectional
Pattern.

Figure 4.4 Spaced Pair.

distance to the source. In addition to the single microphone technique, there are a number of stereo microphone techniques that can be used effectively to capture both the source and the surrounding ambience.

Spaced pair

The spaced pair technique uses two microphones that are spaced apart (Figure 4.4). This captures the source from two

Figure 4.5 Blumlein
Configuration.

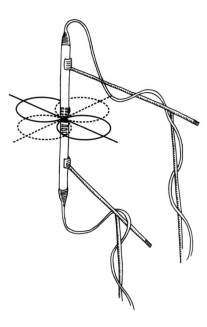

different sides and can capture a very interesting and realistic image of the source.

Blumlein

The Blumlein technique uses two bidirectional microphones on top of each other to capture the sound source in a realistic manner (Figure 4.5). This technique also captures a similar image of the sound behind the microphones. This technique gets used quite often in classical music recording and can be used very effectively while recording stereo samples.

MS

The MS (Mid/Side) technique uses a bidirectional and a uni-directional microphone to capture a realistic sound image (Figure 4.6). Ambient sound is picked up from the sides by the bidirectional microphone, while the unidirectional micro-phone picks up sound from the front. An omnidirectional microphone may be used in the Mid position instead of the

Figure 4.6 M/S Configuration.

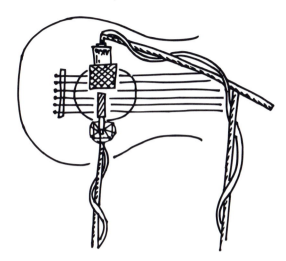

unidirectional microphone if more ambience is desired. This technique requires an MS decoder, which is readily available as a hardware or software unit.

ORTF

The ORTF (Office de Radiodiffusion-Television Française) technique was created by the French national broadcasting system and uses two unidirectional microphones that are spaced 17 cm apart and angled at 110 degrees (Figure 4.7). This is similar to a person's ears. Since this uses unidirectional microphones, you would place them farther away from the source in order to capture ambience as well as the source. Another variation on this is to use two omnidirectional microphones with a divider between them. This creates a similar result. You can also find head-shaped microphone mounts (dummy head) with actual ear-shaped ports on the sides that contain microphones. This is designed to reproduce human hearing.

X/Y

The X/Y technique uses two unidirectional microphones placed right on top of each other (Figure 4.8). Just as with the

Figure 4.7 ORTF Configuration.

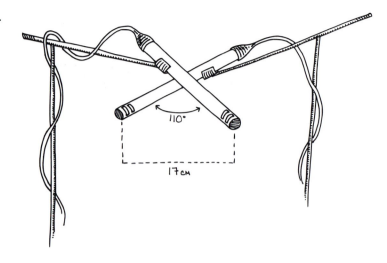

Figure 4.8 X/Y Configuration.

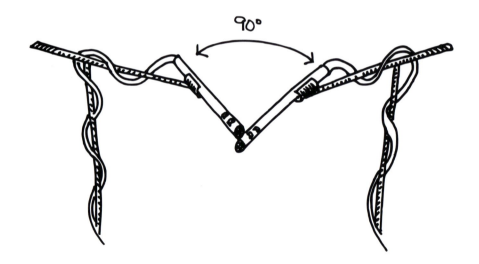

ORTF technique, this should be used only to capture a source and ambience from a distance.

A mix of close and distant microphone techniques

Some situations may require the use of both close and distant microphones in order to capture the appropriate sound of the source and the ambience. Consider, for example, an acoustic guitar in a concert hall. If you were to record the guitar using only distant microphone techniques, the guitar might sound too distant for the desired result. But by placing one microphone up close and a stereo pair at a distance, once they are mixed together, you will have a sound that has both presence and depth. If you want more ambient sound, then you can either reduce the level of the close microphone or raise the level of the distant microphones. The opposite is true if you want less ambient sound. The specific close microphone techniques will be discussed further in the next section.

4.3.3 *Eliminating the ambience*

All of the microphone types can be used when trying to eliminate the ambience of a particular recording space in a recording. The way they are used to accomplish this is quite a bit different from the configuration used to capture both the source and ambience. The primary microphone type you should use for this is the unidirectional microphone. This type is the best at rejecting sounds from all directions except one. This is very useful for recording the direct sound source while reducing ambient sound from around the sound source. The specific types of unidirectional microphones are cardioid, hypercardioid, and supercardioid. The cardioid pattern is the most forgiving of these three, but it has the poorest rejection. Hypercardioid rejects better from the sides but has less efficient rejection from the back. Supercardioid is similar to hypercardioid but is more extreme. An example of a supercardioid microphone is the shotgun microphone, which is typically used in film production (Figure 4.9). Shotgun microphones are often covered in a furry windshield when being used on location.

The other microphone type that should be considered, but is the least musical of the microphone choices, is the parabolic microphone (Figure 4.10). This microphone is one of the best at rejecting ambient sound, but it does so at the price of a less-than-full frequency range.

95

Figure 4.9 Shotgun
Microphone.

Figure 4.10 Parabolic
Microphone.

Any microphone that rejects sound from different directions alters the characteristics of the sounds that manage to come from those directions. Directional microphones rarely have perfect rejection, which means that some sounds will inevitably leak into the microphone from all directions. The basic cardioid microphone might have the least rejection among the unidirectional types, but its leakage will sound less altered and more musical. To really understand this, you will need to test it out and hear the differences.

Microphone rejection sound files

Another key characteristic of unidirectional microphones is the boost of bass frequencies when placed close to a sound source. This is called the *proximity effect*, and it can be a good tool, especially if more bass is needed. But it can also be a distraction when there is no need for more bass. This typically results in more pops from sounds that have a sharp attack or that have bursts of air that are part of the sound. In the recording phase, there is little you can do beside trying other microphone placements, using a pop filter (which works but not perfectly), and using the low-frequency roll-off on the microphone (see below), if available.

Moving the microphone a little to the side of the source sometimes help with the pops. Move it just enough to let the puffs of air bypass the microphone. Make sure the sound from the source is not affected in the recording by this change. If it is, then try using one of the other solutions. You might also try moving the microphone back a little bit; the added distance might reduce the pops just enough.

There are two items that are commonly called *pop filters*. Only one of them is a pop filter, while the other is technically a moisture barrier. A *moisture barrier* is basically a hoop with a fine nylon material covering. A moisture barrier is simple enough that you could make your own from a wire coat hanger and nylon material. A pop filter is either a foam microphone cover or a circular metal grate that helps block puffs of air from reaching the microphone diaphragm (Figure 4.11). Both filters and barriers affect some of the higher fre-

Figure 4.11 Wind Screen.

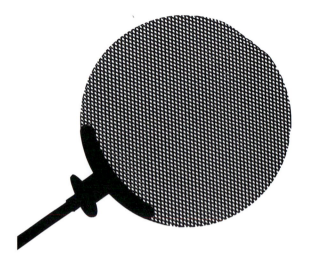

quencies and in some cases are more destructive than helpful. Taping a pencil on the grate of the microphone could also do the same thing by diffusing any puffs of air before they reach the diaphragm. Either way, some pops will make it through a pop filter of any design.

The low-frequency roll-off helps both the proximity effect and the unwanted pops. Pops primarily consist of low frequencies, so setting this option on the microphone might help reduce them (Figure 4.12). If the source has a lot of desirable bass in it, however, then setting this option should be reserved as a last resort. Once the bass frequencies are removed, it can be very difficult and sometimes impossible to get them back. In the case of a bass boost due to the proximity effect, the bass roll-off might work just right to counteract the boost, resulting in a natural sound. This depends on the source, the microphone, and the microphone placement. Check each specific microphone to see how steep the roll-off is, because it may vary from microphone to microphone. The roll-off is a high pass filter. A high pass filter lets the high frequencies pass unaltered, while attenuating the lower frequencies. High pass filters are measured in terms of volume reduction over the period of an octave. Such a filter might reduce by 6 dB

Figure 4.12 Filter Switch.

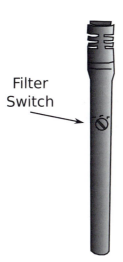

Filter
Switch

per octave or 12 dB per octave. The 12 dB per octave filter is steeper. It is possible to create very steep filters, but the steeper the filters, the less musical the sounds.

Bidirectional microphones can also be used to remove ambient sound. While bidirectional microphones pick up sound from two directions, in certain situations this might be acceptable. If there is still too much ambience, then consider placing a baffle on the other side of the microphone to prevent ambient sounds from reaching the microphone. If this is not possible, then you might be able to move the microphone closer to the source instead, effectively increasing the direct-to-ambient sound ratio. A bidirectional microphone is also an effective choice when recording a source with multiple parts. A set of drums is a good example of this. If there are two drums that are side by side, consider placing a bidirectional microphone between them. The drums will be picked up well, and the ambience coming from the side will be reduced.

Omnidirectional microphones generally have the best full-frequency response and the least coloration. If leakage from other instruments and ambience is not an issue, consider using an omnidirectional microphone first. Omidirectional microphones can be used in a close microphone position setup or when using extensive acoustic baffling. Omnidirectional microphones are not usually the appropriate choice when isolation is desired.

4.3.4 Microphone placements for eliminating ambience

In this section, we focus on more situational microphone placement. The sound and the ambience will determine the placements. There are several guidelines that should be helpful in determining appropriate placement.

If you are using a single microphone, placement should be fairly straightforward. In the case of the unidirectional microphone, place the microphone pointing at the source and away from everything else. Move it around until the sound is what you want.

For multiple microphones on a single source, keep in mind that when the microphones are mixed together, they might not sound very good. Constructive and destructive interference can occur when multiple microphones are used. When this happens, it changes the resulting sound. Even if you listen to each of the microphone feeds separately and each one sounds good, you cannot be sure that they will still sound good when combined. Listen to the microphones both separately and together; this will ensure an end result that works well.

You might want to use multiple microphones on the same source when there are multiple parts to the source. Consider, for example, a drum kit with a number of different drums. In many styles of music, the sound of close miking is preferred, with some overall ambience mixed in as well. Once the drum kit is recorded, each of the individual microphones can be mixed into the ambient microphone's stereo image. For example, the tom drum microphones can be placed left, center, and right in the stereo image.

Another approach with multiple microphones is to take two or more microphones and place them on the same axis. One microphone is close and the others are placed progressively farther away in a straight line. This gives the recorded sound added depth without using the other common stereo techniques.

When creating sampled instruments, you will find that one or two microphones are often enough. Even with instruments such as a drum kit, the samples are triggered individually and so they will be recorded individually. This means that one or two microphones can be used for every component during the recording phase.

4.3.5 *Surround-sound possibilities*

Sampling is one of the last areas of music technology to enter the surround-sound arena. For a number of years, samplers have had multiple outputs that allow intricate output routing, but samplers haven't been used very often to trigger surround samples. The miking techniques for surround-sound recording are a mix of the ambient and close-miking techniques listed above, but with several additional considerations.

The two primary paradigms for surround recording are *performer/audience* and *middle of the band*. The first refers to re-creating the source from an audience perspective, with the source in the front and ambience coming from the sides and back. The middle-of-the-band method puts the listener next to or in the middle of the sound source, with the ambience coming from all directions.

Samplers can utilize either one of these surround mixing styles, but the performer/audience is probably the easiest to implement and the most relevant for sampler uses. The way to create a surround recording for this style is to set up the microphones with three microphones facing forward and two facing backward. The standard surround speaker con-

101

Figure 4.13 Surround Setup.

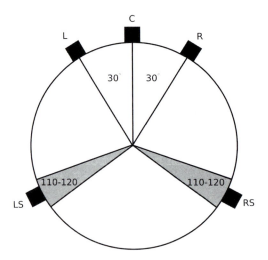

figuration for surround playback has the front speakers spread along a circle, with the center speaker straight ahead and each of the flanking speakers 30 degrees to the left and right (Figure 4.13). The rear speakers are placed approximately 110 degrees from the front on either side.

The standard placement of the speakers can help determine possible microphone placements, because you'll want to capture the sound in a manner that will play back accurately on correctly positioned speakers. In order to record a source that plays back correctly in surround, you should set the microphones in the same place along the standard surround circle but facing the opposite direction from the speakers. The three front microphones should face away from the center toward the front, and the two back microphones should face the rear. These microphones record the source and the ambience throughout the recording space. Once you have the five files from the recording, you edit them the same as you would a single file, but you need to make sure that all the files are edited at the exact same points. Keeping the files the exact same lengths allows them to be triggered properly from the sampler. Each file is assigned to different outputs of the sampler, and these should be routed through your audio mixer or surround monitor system to each of the different speakers/power amps.

If you want to use samples that follow the middle-of-the-band style, you can use any good-sounding microphone placement, including combinations of close and distant techniques. The surround mixing and preparation will have to take place in a digital audio workstation that is capable of surround mixing, because you will be creating a surround mix of the source from each of the different microphones used in the recording. You can creatively place the sound source around the surround image without attempting to create a realistic replica of the source. There are no rules for this situation, and you can experiment with the sounds in many different ways.

The primary limitations involved with working in surround sound are the availability of a full set of speakers to mix on, the number of microphones used to record, and the software with surround capabilities. A great resource to find out more about these fundamental requirements is *Surround Sound: Up and Running*, by Tomlinson Holman (Focal Press, 2007).

Example recording session video

4.4 The recording session

Recording sessions normally run in three phases. These phases are preparation, recording, and cleanup. Each of these phases is critical to the success of a recording session.

4.4.1 *Preparation*

In the preparation phase, you should be able to cover all your needs by asking yourself the following questions:

How?

- Do you have access to the sound source?
- Do you have the computer/sampler?
- Is the computer loaded with the necessary software?
- Do you have the appropriate audio interface?

103

- Do you have the appropriate microphone(s)?
- Do you have all of the necessary cables?
- Do you have speakers (with amp if needed) and/or headphones?
- Are there any miscellaneous items that you need?

Where?

- If the source is not movable, do you have access to the site where it is located?
- If the source is movable, do you have a place selected for the recording?
- Is your recording setup portable?

When?

If the location is a recording studio:

- Has the time been booked?
- Have the musicians been scheduled?

If the location is at home:

- Are there times when extra noise will prevent the session?
- Are there noisy neighbors?
- Is there a "big game" on TV upstairs or people trying to sleep?

If the location is outside:

- Are there times when people are around who should be avoided?
- Is there a big storm coming through or other weather considerations?
- Are there too many nature sounds (like crickets at night)?

If the recording is somewhere else:

- Have you checked open hours for public places?
- Have you checked the lawfulness of using the location?

4.4.2 Setup for session

Once you are in the right place at the right time with the right equipment, you need to take some time to actually set things up properly. Nothing will end a session faster than having to troubleshoot once the time to record has come.

The key to a successful session is to double-check every variable. Test the microphone cables and headphones, because these often have problems that can be hard to track down under pressure. Test the microphones and audio interfaces. If possible, have an extra of each component on hand. If you are recording at home or somewhere else with unlimited access, then having an extra laptop might not be so critical. If you are capturing a once-in-a-lifetime event or something that will require a single take, then having a backup plan in place is critical.

Once everything is hooked up and you are prepared to record, be sure to record a few test takes and then play back the files. Listen for unwanted distortion, buzzes, hum, and anything that might be a problem later. There are many things that can be fixed in an audio editor, but some things are permanently damaging and cannot be repaired.

Once everything is prepared, you should start placing microphones and setting the appropriate levels. It's okay to try moving things around and to take test recordings to ensure that the you're getting the sound you want. Appropriate levels vary depending on the source. The primary goal is to prevent clipping. Clipping takes place when the level entering the microphone or audio interface is too loud and becomes distorted. Another problem is when levels are too low. With digital recording, it is more important to avoid clipping than to avoid low levels, so be conservative in your level setting. Plan for the loudest sound possible. A critical ear at this stage can save a lot of time later in the process.

While setting levels, you should keep everything as consistent as possible and use sound level meters to keep track of the volume. One of the best practices when recording samples is to set one level at the beginning so that when you are editing the recordings for the creation of the different velocity layers, most of the settings will already be prepared.

Velocity layers allow sampled instruments to sound more realistic. When a touch-sensitive keyboard triggers a sample, several things might happen. If the sampled instrument has only one velocity layer, then the same sample will be triggered no matter how hard or soft the note is played. The only difference will be the volume of the sample. If the key is played hard, then the volume will be louder.

If the sampled instrument has multiple velocity layers, then when the note is played hard, a specific sample is triggered that is different from the sample that would be triggered if the note were played more lightly. The different samples are normally recordings of the original source being played louder or softer. The difference is often extreme, because when a source is played louder, the attack sounds different than it would if played softer, and the harmonic content is different at different volume levels of performance. The multiple velocity layers help re-create the original source in a much more detailed way.

This is important during the recording phase because you will have to record each of these velocity layers' samples separately. It is common for highly realistic sampled instruments to have nearly 100 different velocity layers, and it is possible to have up to 127 different layers. However, it is more common to have between 5 and 10 different layers.

Sampled instruments can also have alternate samples that will rotate as the samples are triggered. This helps with realism but means that alternate samples must be recorded. The recording phase is the most critical phase because it is the foundation of the entire process. It is difficult to improve the samples or the instrument design if you failed to record them properly.

4.4.3 Have a plan in place

Understanding the goal behind the sample creation helps to focus the recording session. If you are creating your first sample and trying to master your sampler, then microphone placement might take a back seat in the process. If you are trying to create a sampled instrument that is a replica of a real instrument, then you will have a lot more to plan out and a lot more to keep track of while recording.

Have a note-taking system in place. When creating a complex instrument, you will be doing a lot of recording. Taking detailed notes may slow down the session a little bit, but it will save much time later when editing and mapping the samples. Most audio editing systems have features that work well for this, but many hardware samplers are not as accommodating, and you might want to have a separate method for keeping notes.

Expectation chart template

A detailed expectation chart is recommended. This is a chart that lists all of the items you want to accomplish in the session, with a place to mark when each is accomplished.

This is very useful for a number of reasons. It prevents anything from being overlooked or forgotten. It helps prioritize the recording session. If you are halfway through your allotted time but only a third of the way through the chart, then you need to regroup and adjust how quickly you are working. It also helps from session to session. Make sure the chart is up to date when you finish the session; then at the beginning of the next session, you will be able to get right down to whatever still needs to be done.

4.4.4 File management

Many samplers and audio editing systems have file management features built into them. Read the manual and learn what these features are. In the digital world, managing your files is critical. Keep all your recorded files in the same place. Label them in a way that you will understand later. The name should include descriptive information, such as sound source name and a pitch identifier. For example, a file named *trumpet_C3* would indicate that it is a trumpet playing a C in the third octave. The number with the pitch is a standard system and is the same with most samplers. You could also

107

include other information, such as the sample type. If the trumpet sample is a loud note with a staccato attack, you might label it *Trumpet_C3_s_ff*, indicating a trumpet playing a C3, in a staccato style, and at a fortissimo level. The more information you include, the easier it will be to assemble everything later.

4.4.5 Be critical of everything

Do not waste time by recording something that you do not want. The traditional method of running a recording session is still the most reliable. This model includes a performer, an engineer, and a producer. The performer should not have to worry about the technology involved with the recording session, because when there is a problem, the performer shouldn't be distracted. The engineer is the one who takes care of the technical issues and makes sure that everything runs smoothly. The engineer can listen for noise, clicks, and pops, and other technical things. The engineer also sets levels properly and pushes the record button. Other tasks, such as file management and setup, fall under the engineer's responsibilities as well. The producer should be able to focus on the end goals. Are the recordings consistent enough from one to another? Is the performance achieving the established goal? The producer is often the primary note taker. One of the most important jobs of the producer is to keep the session focused and on task.

Wouldn't it be great if in your sampling recording session, you could afford to have a performer, engineer, and producer? In some cases, you might think it worthwhile to have each of these complementary roles filled by different individuals, but often you will fill all three roles yourself simultaneously. In this case, it is ideal if you can consciously switch between them. Be the engineer as the session is set up. Once the recording starts, switch to the performer and focus on getting the performance you need. When you listen back to the recording, listen like a producer would. If there are technical problems, then switch to the role of engineer. If you make an effort to think in these roles, you will be able to multitask appropriately and create an efficient working environment. The more you do this, the easier it will become.

4.5 Tidying up

Once the recording is finished, there are a few things to do before the session is officially over. Besides the appropriate equipment cleanup, there are also a few other tasks that will help as you prepare either for another session or for editing the recordings.

One of the most important cleanup tasks is backing up all the recordings. This is easiest to do with a computer-based system but is always possible with modern hardware samplers. For the specifics on the hardware samplers, you should check each sampler's individual manual. Typically this involves copying the files to another location so that two or more instances of the files exist. This can be a copy on another hard drive or on a CD-R/DVD-R. When using optical media such as a CD-R or DVD-R, never assume that all are equal. Not only should you buy high-quality blank media, but you should also verify that the burning process was successful. Some hardware samplers allow you to make a backup of your data using MIDI dumps or dumps to a SCSI drive. Either way, these techniques are antiquated and can be a very slow process. However, it is still completely worthwhile to preserve your hard work.

Backup systems

Another thing you need to do after the recording is finished is update all session notes and double-check your file names and notations. Take some special notes about the session if you plan to have another session to continue the recording you have been doing. Take note of any special circumstances that may need to be re-created later. You can also make note of any specific problems that occurred that might be avoided later. Bring a digital camera to take visual notes of session setup information and microphone/sound source placements. Measuring the distances involved with a tape measure

109

is also a good idea if you will have to set everything up again for a different session. It might also be useful to have thermometer and a moisture meter to keep track of the temperature and humidity of the location. These factors can change the sound of the source drastically and should not be ignored. The primary goal is to document as much as possible and to account for every possible situation that may arise.

Once the recording phase is over, you will begin the meticulous instrument creation phase. Having detailed notes and reliable backups of all files is critical for successful project completion.

SAMPLE EDITING   5

Audio files that are recorded need to be edited before they can be properly used in a sampler. While the basic focus of editing is splitting and trimming the audio files, it also includes other maintenance tasks, such as adjusting the volume, adjusting the pitch, adding effects, and putting the sample in the appropriate file format. Editing is also used to help compensate for poor performance and poor recordings. However, it is recommended that the performance and recording be as good as possible because it will help the editing phase be more efficient. That said, today's technology is continually finding new ways to "fix" things after the recording has been completed. Those techniques will be covered here as well, because there are always factors that are out of your control and you will have to use the tools available to overcome any obstacles. In this chapter, the editing process is covered in detail through written explanation and through visual examples. While editing is not overly complex, it is crucial to the sampling process.

5.1   Editing styles

Let's look at a few different editing styles. You should pick the style that best fits how you want to work. Also note that the editing styles are integrated with the recording process. Editing styles are split into the following general categories:

1. Independent editing style
2. Integrated editing style
3. Combination editing style

5.1.1 Independent editing style

Editing the samples independently of the recording phase is often a double-edged sword. If you split the editing completely from the recording, then a more detailed note-taking process is required in order to be efficient. However, when you are creating a highly realistic sampled instrument that has numerous velocity layers, it is often too much work to record and edit in one phase. This is particularly relevant if you are renting a studio or paying a musician to play the source. In this case, it is less efficient to record and then edit every sample in the same phase. By contrast, if you are recording in a home studio and you are playing the source yourself, you might not need to have an independent editing phase.

5.1.2 Integrated editing style

The integrated editing style can be very rewarding because you get to hear the formation of the sampled instrument very early on in the creation process. This style provides immediate quality control, and any potential issues should surface early on. This style works best when you are creating simpler sampled instruments and in situations where the recording phase can be unlimited in time. In a studio where money is being spent, it is typically less desirable to record a note, spend time editing the note, and then spend more time exporting it while the musician waits.

5.1.3 Combination editing style

Combination editing is probably the most efficient editing style. This involves some integrated editing up front, with the majority of editing taking place independently after the recordings are complete. This allows you to ensure that everything will work as planned, and once that is established, the rest of the session can continue with confidence.

5.2 Basic editing

The first place you should look for basic editing techniques is your hardware/software instruction manual. While every

Video examples

system has fundamental similarities, there are enough differences to warrant some personal research. This next section looks at editing in generic terms in regard to samplers and digital audio workstations.

5.2.1 Editing in the sampler

Nearly all samplers have basic editing capabilities (Figure 5.1) but are typically limited to sample start and end point adjustments, pitch changes, and volume changes. There are samplers that are integrated into digital audio workstations

Figure 5.1 Basic Editing.

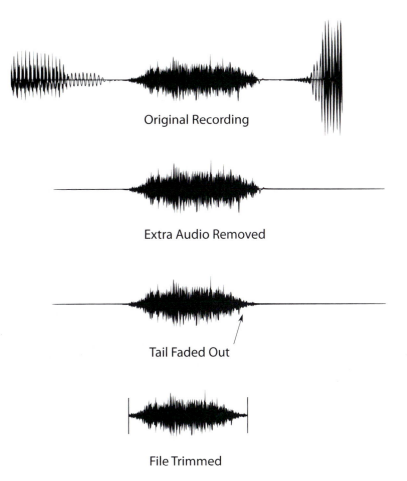

Original Recording

Extra Audio Removed

Tail Faded Out

File Trimmed

and are therefore linked to more capable editing tools, but this is not the case with stand-alone plug-ins and hardware samplers.

Setting the start and end points for your sample is the first step. A recorded audio file may have extra audio before or after the section that you wish to use as the sample. Trimming this audio prevents any unwanted audio from being triggered by the sampler. A graphic display may be included to help you see the waveform, but this is not a critical feature.

A graphic display helps save time by allowing you to visually find the beginning and the end of the audio information. In either case, listening is the ultimate test because some elements of the sound may not be visible in the waveform display. For example, the decay of a sound that is much quieter than the start of the sound is not visible graphically. In fact, only the top 30 dB is visible when using graphical displays. If you edit a sample by sight alone, there is a good chance that you will cut off the sample before the sound actually finishes (Figure 5.2).

The sampler might have tabs that you can drag along the waveform to set the start and end points. Most samplers do not let you fade in and out at these points, so you should listen to make sure there are no clicks and pops when the sample is played back. The way to fix a pop is to move the start or end point until the noise is no longer there. Clicks and pops are caused by the sound starting at a point along

Figure 5.2 Clipping the End.

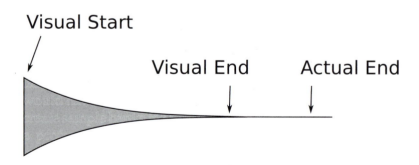

Figure 5.3 Zero Crossing Point.

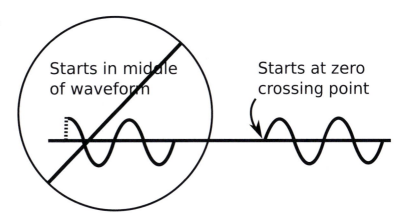

the waveform that is not at a zero crossing, or by too much movement in the waveform in too short a time. *Zero crossing* refers to a point when the waveform is at zero amplitude, or no volume. Graphically, this means the wave is at the center. Editing audio at the zero crossing (Figure 5.3) helps prevent clicks and pops because the audio starts or stops at a point of silence.

If there is a graphic display when editing, you can zoom in and place the starting point at a place where the waveform is crossing the zero point. If this still does not help, then move the starting point farther into or out of the sound, keeping the point at a zero crossing. If, after a period of trial and error, there is still a pop, then the sound may have a pop embedded and other processing may be required (see discussion of fades below).

If the sound requires a pitch adjustment, try using the built-in pitch adjusters. Pitch adjusters on samplers change the pitch of the entire sound in cents or semitones. This shifting is most often accomplished by changing the speed of the sound. A good way to adjust the pitch appropriately is to have the sound you are editing loaded at the same time as another instrument with the desired tuning system. This way you can compare the pitch between the two and make adjustments until the pitch is the same. If the sound gradually shifts out

Figure 5.4 Normalization.

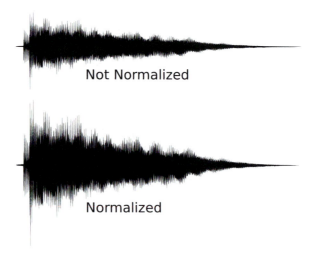

Not Normalized

Normalized

of tune over time, then this type of pitch shifting will not fix the problem and an external pitch corrector is required.

Simple volume changes are possible in samplers as well, which allows you to keep consistent volume levels from sample to sample. There might be a normalization process available that raises the level until it is as loud as possible without clipping (Figure 5.4), but there is always a simple gain change setting that allows you to make the sound louder or softer. Keep in mind that this affects the base file setting and that the sample will be played back at a variety of levels, depending on the velocity of the trigger source.

As far as basic editing in samplers, that is it. Samplers normally have enough features to make functional samples, but the more in-depth editing is left to digital audio workstations and more advanced hardware samplers.

5.3 Advanced editing in the digital audio workstation

DAWs are very complex software applications with many different tools. There are too many specifics to cover in a chapter like this, but there are some standard features that are important to introduce. While the primary focus of this section is on DAWs, hardware samplers have similar features.

The first is non-linear editing. Non-linear editing allows you to change parts of a sound and move them around in any way you want without affecting the original sound file. This is useful when working with multiple recordings of the sound source. If the beginning of one recording is good and the end of another one is good, then you can possibly combine the two to make one new sound file.

You can also take multiple recordings and combine them creatively to make a new combination. An example of this is the creation of a new percussion instrument. The primary sound could be a snare drum, but you could add the sound of a shaker and the crack of a whip (Figure 5.5). These additional sounds may be mixed at a lower level into the original snare to enhance the snare or, at a louder level, to completely change the character of the snare.

Another feature of DAWs is the fade. A *fade* is a decrease or increase in level to or from a zero level. Fades (Figure 5.6) can be used at the beginning and end of audio files to change the attack and release or as a cross-fade when combining two or more sounds together to smooth and camouflage the edit

Figure 5.5 Layering Sounds.

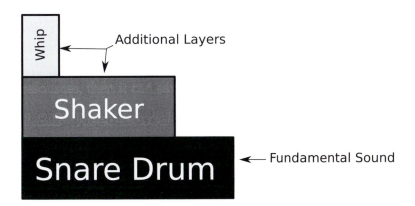

Figure 5.6 Fades.

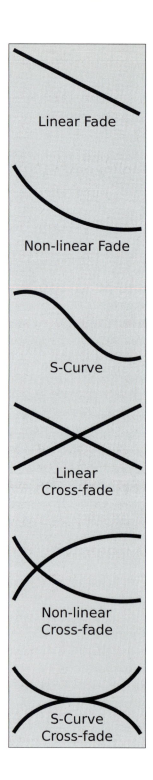

Linear Fade

Non-linear Fade

S-Curve

Linear
Cross-fade

Non-linear
Cross-fade

S-Curve
Cross-fade

points. As an extension of the earlier discussion of pops at the beginning of audio files, a fade can be used to smooth such problems out. Trim a portion of the beginning of the file, and then use a fade to give the file a new attack. You can match the original attack with a fade, or you can use a completely different attack. Fades come in a variety of styles, and you have to test them out to find the appropriate choice. The basic fades are linear, exponential, and s-curve. Linear fades use a steady rate of change, while the exponential and s-curve fades use a non-linear curve. When cross-fading, or fading two audio files together, you can use an equal power (two non-linear fades) or equal gain (two linear fades). The non-linear curves usually provide a less sterile fade with more organic results. No matter which fade type you use, you have to make adjustments until the fade is a transparent transition. For a complete breakdown of fades with audio/visual examples, see the website.

Advanced fades

Most DAWs have pitch shifting capabilities or plug-in protocols to allow third-party plug-ins to shift the pitch. These pitch shifters often have the ability to automatically tune the sound and also allow you to make specific pitch adjustments using a graphical display or other method. The pitch shifting capabilities that you can get in DAWs or that are part of third-party plug-ins are typically better than those that come with the various samplers. Even though samplers are continuously incorporating better and better pitch shifting algorithms, third-party and DAW pitch shifters have consistently outperformed them. See the subsection on pitch correction later in this chapter (Section 5.4.2) for more information.

Audio examples

5.4 Adding extra effects/processors

Effects and *processors* are two terms that are often used interchangeably and sometimes together. A guitar pedal, for instance, is sometimes referred to as an *effects processor*. For clarity in this chapter, these terms are used in a specific way. An effect is something that is added to the original sound, and a processor is something that changes the original sound.

DAWs and stand-alone hardware samplers have a lot of processing and effects tools that vary from system to system but that can be used in nearly every situation to achieve the results you desire. These tools are split into the following general and often overlapping categories.

1. Time effects
2. Frequency processors
3. Dynamic processors

5.4.1 Time effects

The following effects are primarily time effects. However, some of them are closely related to the other categories. These effects delay the original sound, add copies of the sound, or add multiple reflections to the original sound. These effects develop over time and often carry on past the end of the original instance of the sound.

Delay

The time effect used most often, both independently and as a part of other effects and processors, is the delay effect. A *delay* holds back a sound in time and releases it at a specified interval. Typically, the original sound remains unaffected, while a copy is delayed and mixed in with the original. The copy may also be part of a feedback loop that creates multiple copies that are mixed in with the original. These copies are spaced apart and can be timed to any desired tempo.

The delay effect is used for several purposes. The first is to add space around the sound. A delay can emulate how sound reflects off different surfaces. This emulation can give a dry sound additional depth and realism, placing the sound in a pseudo-reverberant field. A delay effect can also be used to change the sound and give it a larger-than-life quality. The result is determined by the delay time and the feedback amount. A short delay with very little feedback creates a slap delay and gives the sound the illusion of a larger sound. A longer delay with a longer feedback adds space around the sound. In either case, experimentation is required to find the sound you are looking for.

If the sound source you are recording has a tempo (like a drum loop), then you might consider using a timed delay that matches the tempo of the source. This helps the delay blend in and can be a pleasing effect. If you want the delay to stand out, then purposely mistiming the delay can accomplish this. A well-timed delay can alter the sound or simply enhance it.

Delays are also used on other effects. As you look through all of the other effects and processors, you should note which of them have a delay component and see how it is related to the delay discussed here. The use of delay in so many other areas demonstrates that the effects/processor categories have much overlap.

Various delay-based effects

This section focuses on the principle behind a group of effects that use delay to obtain their results. These include chorus, flanging, and phasing. Each of these uses a copied and delayed version of the original signal that is then mixed in with the original to create interesting effects.

When a sound and a copy of a sound are mixed together, they add together and the end result sounds louder. When one of them is delayed, interesting things begin to happen because of phase cancellation and boosting. The delay can even be set to follow an LFO (low-frequency oscillator), which changes the delay based on a cycling oscillator and can create interesting pitch variances.

These types of effects are almost always used as creative effects. Sometimes, instead of pitch shifting a sound, you might use one of these to camouflage pitch problems.

Reverb

One of the most often used time effects is reverb. This effect emulates acoustic reflections that are added to the sound as a way of adding dimension. Reverb units have a set of standard parameters that relate to specific acoustic equivalents. These parameters include the direct sound from the source, the initial reflection, other early reflections, general reverb, and decay time (Figure 5.7). A reverb unit attempts to re-create this process using a complex series of repetitions of the sound source. Most reverb units have a basic algorithm that defines the fundamental space of the emulation. This might be a specific room, a cathedral, or a concert hall. The pre-delay setting defines the time between the original sound and the first reflection. Increasing the time on this setting increases the perceived size of the room. There is also a decay setting that lengthens or shortens general reverb time. Other settings such as equalizers and modulators vary from reverb to reverb, but most reverbs do have additional and useful features.

Another way of setting reverb parameters is based on convolution. This type of reverb creates highly realistic replications

Figure 5.7 Reverb.

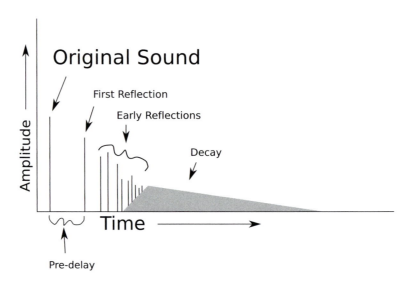

of specific locations, because the source material of the unit comes from those specific places. A set of microphones and a speaker are set up in a location that is to be captured. A sweeping sine wave is broadcast through the speaker and is captured by the microphones. The microphones capture the sound of the sine wave in the room. The resulting sound file can be decoded using the recorded sweeping sine wave from the room and the original sweeping sine wave. The sine wave is effectively removed, and a file called an *impulse response* is left which can be used in convolution reverb units instead of a more traditional algorithm to create the basic sound. An alternative method of capturing an impulse response uses a short sound burst, commonly a balloon burst or starter pistol. While these are easier to use, they provide less accurate results.

Reverbs can be used in much the same way as the delay effect discussed earlier. They can be used to create a space around a sound or the give the illusion that the sound is bigger than it really is. This is accomplished by adjusting the basic reverb parameters/impulse response and the decay time. If the basic reverb is set to a large room but the decay time is set very short, then the source will sound larger without sounding like it is in a large room. If the decay time is lengthened, then it will sound like it is in a big room.

Reverb units also have equalizers built in. An equalizer is a frequency-specific volume control. Using an equalizer can help achieve natural results. The sound passing through a reverb unit is often varied, and the reverb doesn't always suit the sound. An equalizer can be used to smooth out frequencies that do not sound natural or boost frequencies that are lacking. In either case, equalizers are used enough that they are now a part of nearly every reverb unit. See the equalizer subsections in Section 5.4.2 for more information.

Reverb units are used in a multitude of situations, but there are times when they should not be used. It is important to know when those times are. When your sampled instrument is finished and it gets used in a recording session, most likely it will be passed through a reverb as a part of the mix. In this case, it might be desirable to have a dry sample with no reverb or delay. If you use reverb on the sample, then when it is mixed with other reverbs later, either by you or some other end user, it might not sound good. In later mixing sessions, the engineer might have a specific sound in mind that could clash with the reverb on the sampled instrument. On the other hand, if your sampled instrument has a specific sound that is built around a specific reverb, then including the reverb is a requirement. When it is used later, the engineer has to take the reverb sound into consideration. Don't be afraid to add reverb to the source, but if it is not necessary, you might consider leaving it off completely. Another option is to create two versions, one with reverb and one without reverb.

5.4.2 *Frequency processors*

Frequency processors affect the frequency spectrum, affecting the timbre and tone of the audio that is passed through the processor. They are considered processors because the frequency adjustment is made to the original sound, and there are no copies of the original involved. Frequency processors are used very often in music production; likewise, in sample creation. You will find these processors especially useful when recording in less-than-ideal acoustic environments and with less-than-perfect sound sources.

Filters and equalizers

A *filter* is a singular control over the amplitude of a specific frequency band. An *equalizer* (Figure 5.8) (EQ) is a collection of filters. An EQ is a very common processor and one that has a variety of implementations. There are three types of EQ: the parametric EQ, the graphic EQ, and the paragraphic EQ.

The basic filter types used in EQs are the peak filter, the high pass filter, the low pass filter, the band pass filter, and the

Figure 5.8 Equalizer.

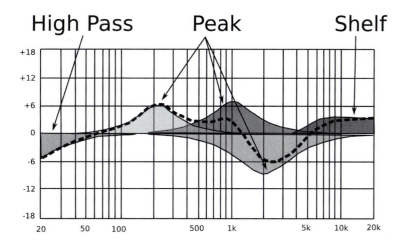

high/low shelf. All of these filter types are building blocks for the other types of EQ. The peak filter boosts and cuts frequencies across a range of frequencies. High/low pass filters cut either the low or high frequencies. Band pass filters are a combination of the high/low pass filters, which results in a band of frequencies that are left in the middle. The shelf filters boost and cut high or low frequencies from a specified point through the rest of the range. The curve looks like a shelf.

The parametric EQ has several bands of filters with adjustable parameters. This EQ typically has a high/low pass filter and several adjustable peak filters. This is one of the most flexible EQs.

The graphic EQ consists of a row of fixed frequency peak filters split into octave or 1/3-octave bands. It is called a graphic EQ because the sliders that control the bands are laid out in a row that graphically shows the curve of the frequencies being affected.

The paragraphic EQ is a mix between the parametric and graphic EQs. It is primarily a parametric EQ that has additional bands.

Nearly every equalizer uses the same parameters to adjust the frequency response. There is gain for every band, a frequency adjustment, a bandwidth adjustment, and a slope (called Q) setting. Not all of these parameters are available for each type of equalizer, but you can expect to see them in a variety of places.

The gain control is very similar to a fader on a mixing console. It can boost and cut the volume of the chosen frequency over the range of the bandwidth. The frequency control selects the frequency to be affected. Most equalizers split the frequencies into ranges so that one band covers a set range and isn't able to move through the entire range. The bandwidth setting determines how wide a range is affected by the gain control. The Q setting is inversely proportional to the bandwidth, and so a small Q translates to a wide bandwidth. The bandwidth can be determined by dividing the center frequency by the Q number, and the bandwidth is defined as the width of the frequency curve in excess of 3 dB up or down from the tolerant window. Most equalizers use a graphic display that provides an intuitive editing interface that is easy to use.

Application of EQ

An equalizer is a powerful tool that can do a lot of things but is often overused or used inappropriately. Listed below are some general guidelines for EQ use.

Use EQ as a last resort:

Anytime you can avoid using an EQ, you should. An EQ can introduce audio degradation if used too much. If you do not need to use an EQ, then don't. Alternatively, you could:

1. Change the microphone or switch the polar pattern.
    a. Different microphones have different frequency responses.
    b. If you want more bass, try a cardioid pattern and take advantage of the proximity effect.

2. Change the microphone placement.
   a. Moving the microphone away from the source decreases the higher frequencies.
   b. Moving the microphone off axis from the source decreases the higher frequencies.
   c. Moving a directional microphone away from the source removes the proximity effect.
3. Change the sound source's environment.
   a. Different types of acoustic treatment are designed to absorb or reflect different frequencies.

The design of analog filters (emulated by the majority of digital EQs) results in a phase shift that changes with frequency and so can contribute to constructive or destructive harmonic relationships in the final mix. This means that using an EQ not only boosts and cuts frequencies, but it also affects the phase relationships. Alternatively, there are some modern EQs that use an extremely transparent process called *linear phase EQ*. This type of EQ changes the fundamental sound less than other EQs but in doing so requires a much longer delay time as part of its processing. Using a linear phase EQ in real time with samplers is not recommended due to the long delay involved. However, in some cases, this type of EQ may be preferred to traditional EQ types when seeking a transparent sound. In either case, you should use EQ sparingly or only when you have a good reason.

Boosting with an EQ:

If you record a sample and you want to change how the sample sounds, then consider boosting certain frequency ranges with an EQ. Boosting with an EQ can only add more of the sound that is already there. If there is some bass but you would like more, then boost the bass using an EQ. If the sound is dull, then boost the higher frequencies. Boosting can change the sound either slightly or drastically, depending on the sound's original spectrum and where you are boosting. If you are working with a distorted guitar, for example, boosting the sound at a frequency where there is a lot of distortion can really bring out the edginess of the sound. A slightly boring sound can become alive and exciting.

Cutting with an EQ:

Cutting a frequency range a small amount will not change the sound as much as boosting does, but it can make a big difference. When cutting smaller ranges, you can clean up the sound and remove unwanted frequencies. Let's say you are recording an acoustic guitar and the sound is "muddy." Maybe it is the room where the recording was made that is boosting certain frequencies, or perhaps the guitar has a fundamentally "muddy" sound. Cutting some frequencies in the trouble areas can help clean up the sound. If you cut a lot of frequencies, then it is the same as boosting the range that wasn't cut. Think of boosting with EQ as a way to alter the sound source and cutting as a way to simply clean up the sound.

Sweeping to find the right frequency:

If you hear something you want to change with EQ but you are not sure which frequency to change, there is a method to find the proper place. First set a narrow Q setting. Remember that the higher the Q setting number, the narrower the bandwidth. Boost the filter quite a bit and then sweep through the spectrum until you hear the range that you want to change. If you want to cut the frequency, then also try sweeping through with the frequency cut. Most people find it easier to hear the appropriate place by sweeping a boosted frequency, but don't be afraid to try both. Once you find the desired frequency, set the Q and the gain to whatever settings work in the situation.

You cannot EQ what isn't there:

An equalizer does not create new frequencies. If there are no bass frequencies in the original sound, then boosting the bass will not accomplish the goal of having more bass. In this situation, you should use a different tool to accomplish the bass boost. One way to add bass to a sound that doesn't naturally have any bass is to duplicate the track and shift the duplicate down an octave. Once it is shifted down, put it through a high pass filter to remove the higher frequencies and mix the resulting lower-frequency track until it fits into the original.

Use of graphic EQs:

Graphic EQs are mostly used in conjunction with loudspeakers and public address (PA) systems. The primary situation in which you (as a sampler creator) might want to deal with a graphic EQ is if you had to listen to your recording sessions on speakers in a room with acoustic problems. The graphic EQs can be used to help "tune" your system to the room and create a flatter frequency response. Ideally, you would change the room's acoustics before using an EQ to fix things, but an EQ is the natural second choice.

Removing the rumble:

Use a high pass filter to "roll off" any unwanted lower frequencies. If the sound source you are recording has no frequencies below a certain range, then set the microphone roll-off switch, if available (this is a built-in EQ), or do the same thing with a high pass filter. The purpose of this is to prevent some low frequencies that might come from bumping the microphone stand or, if working at home, from household noise such as an air conditioner or furnace. If you are using headphones or speakers that are not full range, then you need to be extra careful with low-frequency monitoring, because those systems do not give you a good idea of what is happening in the lower ranges.

Pitch correction

Pitch shifters and correctors are another example of frequency processors and are used extensively in sampling. If the source is not in tune, a pitch processor can be used to tune it. It would be ideal if the musician could tune the source during the recording phase, but in some cases this doesn't happen. One primary consideration when using a pitch shifter is whether to alter the pitch and time together, or to do it separately. Early pitch shifter designs worked by speeding the audio up or slowing it down. When the audio is sped up, it changes the pitch as well. A digital file, however, can easily be shifted without changing the speed of the file. Nonetheless, changing the speed of a file still makes a good-sounding pitch shift, because all that is required to shift this way is a change in the clocking of the digital file. Shifting a file while keeping the timing unchanged requires more processing and complex algorithms.

One type of pitch processor is a pitch shifter that transposes the entire sound up or down by a set amount. This is the type of processor that samplers use when mapping. Another type is a pitch corrector. This analyzes the pitch of the source and adjusts the pitch to fit in a set scale or tuning system. This is very useful when working with a source that is not tuned to any particular tuning system.

Pitch correction tools often have two operating methods, one automatic and one graphic. The auto setting can be very handy when tuning a lot of samples. But you still need to check the results, because most correctors are not perfect and there is a good chance for audible artifacts. The graphic mode is the most precise and allows you to perform very detailed tuning changes. Typical controls for pitch correctors include tuning time, scale selector, source range selector, master pitch shifter, vibrato, and note selector/bypass.

The tuning time dictates how quickly the corrector will react to a sound source that is out of tune. The corrector compares the source to the chosen scale, and whenever the source is outside the scale, the source is shifted until it fits into the scale. The range selector should be adjusted to match the source material. The possible choices are usually soprano, alto, tenor, bass, and instrument. The corrector uses different algorithms depending on the source selection. The master pitch shifter allows the entire source to be shifted. The vibrato tool can add vibrato to sound. This is used to accentuate already existing vibrato, or to re-create vibrato on a sound that has had the vibrato removed by excessive pitch correction. Note selectors and bypass switches allow certain notes to be left uncorrected. This is useful when only certain notes are out of tune.

Be aware that pitch correction tools still cannot dissect a source that has harmonies or multiple notes played simultaneously. If you are using a guitar chord as a sample source, then it should be carefully tuned before recording because there is nothing that can be done later to fix its tuning problems.

Distortion processors

Distortion processors originated in the realm of electric guitars in the 1960s but have evolved into effective tools for use with a variety of source material in a variety of styles. The original distortion effect was created by "damaging" the guitar amplifier's speaker; this evolved into different methods of overloading the signal to add a variety of harmonic information. Tube-based distortion emphasizes odd harmonics, whereas solid-state distortion emphasizes even harmonics. New digital technology emulates both types, and there are plenty of distortion units that range from specific hardware emulators to creative ventures into new distortion techniques.

Application of distortion processors

Distortion units can be used to create realistic distortion guitar tones by passing a clean guitar sound through and "re-amping" the sound. A true re-amping would consist of recording a guitar straight into a microphone preamp, without adding any further effects or processors, and then sending the recorded sound back out into an actual guitar amplifier.

This allows the greatest amount of flexibility, because you can keep the original performance without committing to a specific sound and you can tweak the sound as much as you like. Sending the signal to a hardware guitar amplifier/processor is one way to do this, but there are software distortion emulators that can be used to re-create that sound fairly realistically.

In addition to using distortion processors to re-create realistic sounds, you can also use them to create interesting sounds that are atypical. Try processing a series of words or phrases through a distortion unit and use them as the foundation for a new instrument. Maybe you would like a distorted oboe sound or a piano that is pushed to the limits. There are no rules and very few guidelines when experimenting with distortion units.

5.4.3 Dynamics processors

Dynamics processors affect the volume of the sound being processed. *Dynamics* refers to changes of volume over time. Sound that has both high volumes and low volumes is said to have a wide dynamic range. It is possible to reduce or increase the dynamic range using dynamic processors. Dynamic processors are similar to the equalizer, which affects the volume of specific frequency ranges of a sound. In fact, multiband compressors (see below) are often used in place of equalizers because they affect the dynamic range of frequency bands. When the goal is dynamic control, then a multiband compressor is typically considered a dynamics processor; but when the goal is frequency control, then it can be considered a frequency processor.

Compressor

A compressor (Figure 5.9) controls volume using adjustable settings and can be considered an automatic fader. In fact, nearly everything a compressor does can theoretically be accomplished by moving a volume fader with your hand. In

Figure 5.9 Compressor.

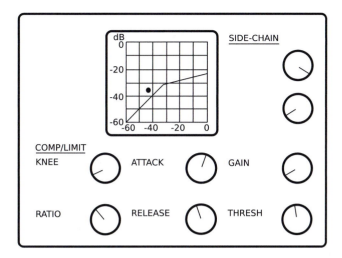

practice, though, a compressor can make adjustments much more quickly and with a much shorter reaction time than a person ever could.

A compressor looks at the volume of the original sound and turns on when the sound reaches a certain threshold. A ratio determines how much the compressor will reduce the level. A 1:1 ratio will do nothing, because for every 1 dB above the threshold, it allows 1 dB to pass through. A 5:1 ratio compresses the audio quite a bit more, because for every 5 dB above the threshold, only 1 dB is allowed through. So, if a sound reaches 10 dB above the set threshold, it is reduced to 2 dB above the threshold.

A compressor also has attack and release times that indicate how quickly the compressor will begin to work once the threshold has been breached and how quickly the compressor will return to inactivity once the source sound falls below the threshold level. Other common compressor settings include the knee and makeup gain. The *knee* refers to the transition point where the threshold turns on the compressor. A hard knee will snap the compressor on, and a soft knee has a gentler transition. *Makeup gain* is a built-in volume control

133

that allows the overall sound to be turned up after it is compressed, taking advantage of the freed-up headroom.

Application of compressors

A compressor can be very useful in a sampling environment. The primary use is to prepare the samples for mapping to different velocity layers. In an ideal world, you would record perfect samples and every single take would be in an exact velocity layer range from the start. But this is not always possible or practical. The other use is to correct the inappropriate results that can occur when a sound source varies in volume over time. A violin sample that is 5 seconds long, for example, might decrease in volume in the middle seconds. This volume change is not a good thing, and a compressor can be used to even out the volume changes.

Most often, adjusting the overall volume of a file suffices when preparing sounds for velocity layers. However, a compressor can provide more accurate and consistent results. When you are preparing a specific velocity layer that has 50 samples that will be mapped over a keyboard, you could adjust each by hand, using a level meter to see the current and changed level of each sound. You could also use a compressor to do the same thing, but with several additional features. Here's how:

Set up a compressor and use a threshold setting just below the velocity you are attempting to achieve. If the sound is still softer than this target, then set the threshold even lower, below the actual sound level. Setting the threshold below the current sound level allows the compressor to alter the dynamic range. If you want a more natural sound, then do not set it too far below the desired velocity range. If a compressor is set too far below the range, it might change the sound of the audio too drastically. A compressor, if set incorrectly, discolors the sound and can even cause distortion.

Next, set the ratio to an appropriate setting. If you are trying to maintain a natural sound, then a lower ratio is best. This might be around 4:1 and below, but the results depend on the source material. If you are having a difficult time hearing what the compressor is doing, set the threshold and ratio to extreme settings, and then you should be able to hear something going on. Once you hear the compressor working, reduce the settings to appropriate levels.

Next, set the attack and release times. For source material with a quick initial transient, like a drum strike, consider that a short attack setting may stifle this desired percussive attack. Try both a long and a short setting, and use the amount that works the best. If you would like the compressor to continue working after the signal has fallen below the threshold, then turn up the release time. This allows the compressor to keep the sound compressed. Consider doing this when there is an obvious audible change in level. If it is distracting, then keeping the compressor engaged longer may help the transition sound smoother from compression to no compression.

Once you are happy with the compression, you can set the makeup gain. You should have a level meter in place so that you can see the exact output level of the compressor. Set the makeup gain so the output matches other samples in the same range. The important thing to do is create a realistic set of samples at various different levels. Later, when mapping the samples into velocity layers, you will be tweaking them until they sound natural when triggered in their velocity layer. When setting the volume levels of samples, there is a certain amount of guesswork. Until the samples are mapped, it is hard to tell how effective the different samples will be.

The exact implementation of the above steps will vary from situation to situation. It depends on the source material and what you are trying to accomplish. The reason to use a compressor is to alter the dynamic range of a group of sounds and give them similar dynamic ranges. If this is not the end result for a particular situation, then maybe a compressor is not the appropriate tool. For samples requiring an extreme level of detail and a natural sound, then using a compressor is often the last thing you should do.

Compression can also be used as an effect or as a creative tool. Sending a copy of the original sound through a compressor and then mixing them back together can achieve a compressed but natural-sounding result. This technique is used with drums and percussion to create an even and consistent sound without completely sacrificing a natural dynamic range.

Overcompression is not always a bad thing either, and it is a type of sound that is prevalent in modern pop/rock styles. Using a compressor to achieve this might be just the thing you are looking for. It might not result in a natural sound, but who says that everything has to sound like the original? Often times, an altered, larger-than-life version is preferable over the original.

Limiter

A limiter is very similar to a compressor except that the ratio is set to at least 20:1. The sound level is typically increased until it reaches the threshold, which is usually set between −5 and 0 dBFS. Limiters effectively shave off the top portion of the dynamic range, allowing for more headroom at the top of the dynamic range. The settings on a limiter are very similar to a compressor, and some compressors can actually function as a limiter if the ratio is set high enough. True limiters are designed to control the dynamic range with very little distortion and coloration, while compressors are not as transparent.

Application of limiters

Limiters are not usually utilized in the sample creation process, but there are several reasons why you might use them. One is to make a sound as loud as you possibly can without simply turning it up until the distortion starts. A limiter is able to turn down the loudest peaks of the audio, just like a compressor, but it is designed to do it even better and with more transparent results. Another use is when you are making a sampled instrument that includes multiple, fully produced and mixed stereo audio files that are at different levels. A limiter is used in this situation to help bring the different files closer to the same level without changing the sound of the files too drastically.

Gate

A gate is another automatic volume control, similar to the compressor. The settings are nearly the same, but when a sound goes below the threshold of a gate, the sound is reduced until the sound returns above the threshold. Most gates allow the sound to be turned down or turned off when the source is below the threshold. In many instances, it is not desirable to let a gate close completely, because it is not desirable to hear the noise floor pump between sounds.

Applications of gates

A gate is used to attenuate sound below a certain threshold. Consider using a gate in situations where there is low-level noise or when you are using multiple microphones and you want to minimize what the different microphones capture from the source.

Recording a guitar through a tube amplifier is a good example of a possible gate application. When the guitar is not playing, there is a certain amount of amplifier noise that you may wish to remove. Instead of editing every single note by hand, you could use a gate to remove or minimize the presence of the noise. The amplifier noise is still there when the gate opens, but it is partially masked by the guitar performance.

Gates are used during music production more often than when creating samples. As you edit and prepare samples, the things that can be accomplished by using a gate are usually done manually, as using a gate is actually less efficient.

Expander

An expander is one of the few tools to undo what a compressor has done. An expander takes the differences in dynamic range and makes them more extreme. What is louder will be turned up, and what is softer will be turned down. There are almost no instances that you would use this in the sample creation process, for the same reasons that a gate isn't typically used. However, you might try it out if you are looking for an interesting effect.

DeEsser

A DeEsser is a mix between a frequency processor and a dynamic processor. It is used to attenuate the "sss" sound—most often on vocals, but also anytime an "sss" sound is too loud. The reason the "sss" sound is sometimes overtly present in vocals and other sources is because that particular sound is created by a gush of wind moving through a narrow corridor (tongue and teeth). The extra air that is required to create the sound accentuates the sound at the microphone and often times seems inappropriately loud. The deEsser is a compressor that affects only the range of frequencies around the "sss" sound. The settings include a threshold, a gain reduction amount, and a frequency adjuster with a limited range. Some deEssers also have a monitor switch that allows you to hear the sound you are removing from the source. This can be useful when trying to find the specific frequency.

Multiband compressor

The multiband compressor is another dynamics processor that is also a frequency processor. It is very similar to the deEsser but with expanded capabilities. Instead of a limited frequency band, it might have up to five or six bands that can be compressed independently of one another. The reason this is placed in the dynamic processor section is because it is more of a compressor than an equalizer. However, it works very much like an automated equalizer as well. Instead of an overall boost or cut in different frequency ranges, it compresses the dynamic range of the specific frequency ranges, providing more headroom in that range. Each range can then be boosted or cut in a controlled manner. You can also leave the overall volume of the ranges relatively untouched, but limit the dynamic range so that it is kept under tighter control.

Application of multiband compressors

The most likely place that you would use a multiband compressor is with sound sources that have out-of-control bass. Using a multiband compressor can help get control of the dynamics of the bass frequencies without affecting the other frequencies. A normal compressor affects the overall sound, which means it will affect the bass as well as the other frequencies. An equalizer can boost or cut the bass, but that will not bring it under more control. By using a multiband compressor, the end result is a bass range that stays consistent and remains at a specific level throughout the sound.

A didgeridoo or other low-frequency instrument might benefit from a multiband compressor because you can tighten up the lower frequencies without changing the overall sound of the upper frequencies. This keeps the sound more natural and accentuates the lowest parts of the sound. You can also use a multiband compressor to control other frequencies that might not be as consistent as you would like. As with all compressors, it is possible to remove the natural dynamics from a sound. Be careful not to overdo it, unless you are trying to get a really squashed sound as a creative effect.

5.4.4 Applying effects and processors

Now that we have covered a few of the basic effects and processors, it is time to discuss how these are applied to the audio files you are working with. There are a number of terms used by different software and hardware platforms. The terms listed here are not universal but the definitions are universal. You should become familiar with the terms used by the specific technology you are using. Some of the following techniques can only be accomplished in DAWs, and if you are using certain hardware samplers, you might be limited in the options available.

- *Real-time plug-ins:* This type of software module is inserted in the audio path. The audio is processed as it passes through, and the processing is not written to the audio file. This method requires CPU processing as long as the plug-in is operational. This is also true for hardware effects and processors; the audio passes through them without being stored somewhere permanently.
- *Bounce to disk:* This is the function that allows the real-time plug-ins to be applied permanently to a copy of the original audio. The audio is played through the plug-ins and then is written to a storage device.
- *Offline Processing:* This type of software module is applied to the audio file, which is then stored. This method requires CPU processing while the module is processing the file and while it is being written to the storage device. The CPU is then free to do other things.
- *Export file:* This allows files to be exported from whatever system you are using. Real-time plug-ins will not be included in this method, but any offline processing will be included.

Both real-time and offline processing can be done with multiple modules, and they can even be used side by side. The key is to have a file at the end that can be imported into the sampler you are using. There are several other factors that are important to consider in this export process, because the end file format should match the sampler's format.

5.4.5 Export settings

Modern samplers allow digital files to be imported at the highest possible settings. The two primary considerations are bit depth and sample rate. *Bit depth* determines the maximum dynamic range, and *sample rate* determines frequency response. For hyper-realistic sampled instruments, you should use 24-bit files. The sample rate is a less critical consideration. as long as it is at least 44.1 kHz, which allows frequencies well above the limit of pitch recognition and also meets the typical professional expectations of a range from 20 Hz to 20,000 Hz. A simple way to remember the frequency range of a specific sample rate is to divide the sample rate by 2 to find the highest allowed frequency. In the case of 44.1 kHz, the highest frequency is 22,050 Hz, which is beyond the hearing of most people. Higher rates can be used when creating high-quality instruments, but memory considerations, both hard drive storage and RAM, might be affected by the sample rate decision. The higher the sample rate and bit depth, the more storage space is required and the greater the memory and CPU consumption. If you are creating a highly detailed instrument with a large number of files, you might consider using a lower sample rate but keeping the higher bit depth. If the dynamic range is not a critical consideration, then consider using a bit depth of 16, because it will free up your resources to have smaller file sizes taking up the power of your system. Some samplers have a 16-bit mode that uses fewer resources without forcing a compromise on the source resolution. When you want to use the samples in a final mix, you can switch back to 24-bit mode and utilize the fullest resolution. With modern computers, you may never have to worry about this compromise in quality because computing speed and memory are always increasing. A good rule is to use the highest quality that is possible and reasonable.

CREATING THE INSTRUMENT   6

Video tutorials

6.1   Importing files into zones

Once your recorded audio files have been edited, the next phase is to import the files into a sampler. *Mapping* is the process of assigning the individual audio samples to specific keys and key ranges. An imported audio file is typically triggered by more than one key and in a velocity range. The audio file and all of the related parameters are collectively called a *zone* (Figure 6.1).

Samplers that have zone grouping capabilities allow multiple zones to be edited at the same time, and it makes changing parameters much easier.

6.1.1   Setting the root note

After the initial import, the next step is to assign additional information to the zone. You need to know the pitch of the imported sample to set the root key in the sampler. If the imported file is a C3, then you will set the root key to C3 (Figure 6.2). This tells the sampler that the note should not be pitch shifted when a C3 is triggering the sample. If an incorrect root is set, then the sample will play in the wrong key in relation to the other samples. If a C3 sample is set with a root note of D3, then when a D3 triggers the sample, a C3 will be played. Some samplers are able to auto detect the root setting when the sample is imported by reading the file name. If the file name is formatted in a specific way, then this system works. Other samplers are able to analyze the pitch of the audio sample and set the root to correspond to the actual pitch.

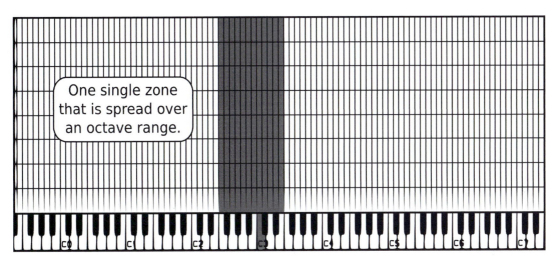

Figure 6.1 Single Zone.

Figure 6.2 Root Note.

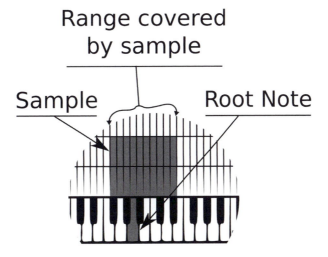

You also need to define how many notes the sample will cover. In the past, samplers did not have enough power to have separate samples for every note. This was due to limitations in processing power, RAM amounts, and hard-drive speeds. Now, it is possible to have separate samples for every note, even though this is not always done. The reasons for not doing it include time, money, and compatibility. Also, it might not help create a more realistic instrument to use samples on every note. The more files a sampler uses, the more time and money it takes to create those files. Companies that sell sampled instruments try to find a good balance between quality and profit. In terms of compatibility, sampled instruments are designed to be used on as many different systems as possible. If you can create a decent sampled instrument using fewer source samples, then it will definitely save money. It will also have a better chance of working on older systems that have limited processing power and memory. Some sampled instruments do have individual files for every note, but this is still the exception and not the rule.

One of the ways that assigning key ranges for zones is accomplished is through a graphic display. A box that is lined up against a piano-style keyboard represents the sample. The root note is indicated when the box is selected. The edges of the box can be dragged to the left or right, allowing you to choose how many notes will trigger the same file (Figure 6.3).

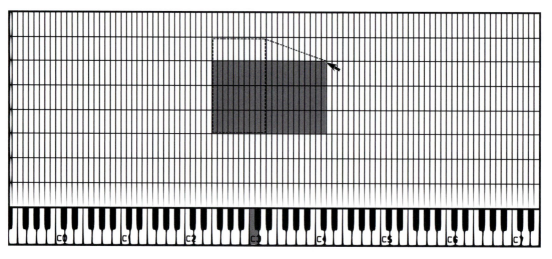

Figure 6.3 Graphic Editing Tools.

Unless the sampler is told not to, it will transpose the file according to the note that triggers the file. If the root is C3 and D3 triggers the file, then it will be transposed up two semitones. This process does not require a graphic display, and most samplers also have a way to do this using text-based parameters. In either case, you need to listen with your ears to verify the results. Never trust that the graphic display will ensure quality results.

In your specific project, there might be a sample for every possible note or there might be only a few samples that are spread out. Obviously, the more samples that are mapped out, the more every note will sound independent from the other notes. If the project is an attempt at a hyper-realistic instrument replica, then more samples are required. If you are creating a much simpler instrument, you can get away with many fewer samples.

Audio examples of instruments with varying amounts of samples

6.1.2 Setting the zone outputs

Another parameter that needs to be checked is the audio output routing. If the sampler is capable of multiple outputs, then you have to set the output for each zone. The default is normally the primary stereo output. If you are creating a surround instrument, you have to import the surround files and set the outputs for each separately.

6.1.3 Zone overlap

Once you have set the zone parameters for the sample, you can import other files into other zones. When two zones overlap, both sounds will be played simultaneously.

This is a powerful tool, because layering the zones directly over each other can create new sound combinations. You can take an ordinary flute sound and add a percussive attack to it (Figure 6.4) or take a drum sound and add the sound of a guitar string as the sound decays.

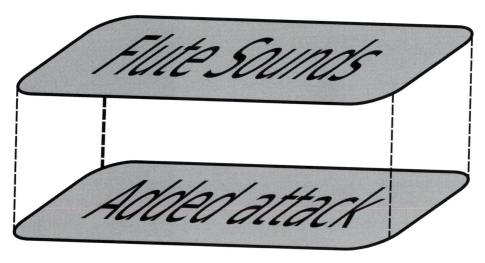

Figure 6.4 Layered Sounds.

Velocity ranges

As a zone is triggered, the sampler takes the velocity of the trigger and uses it to switch between zones. One zone may be triggered when C3 is played in a velocity range of 1 to 60. The MIDI standard uses a range from 1 to 127, and the velocity ranges of zones may cover all or part of that range. A second zone might be layered over the original C3 zone and is triggered from 61 to 127 (Figure 6.5).

These velocity ranges can overlap, and there are cases when this is desirable. A basic sample may be enough to cover the entire 1–127 range, but in the louder velocities, a second sample may be added to change the sound. This is similar to how a real instrument might change as it is played louder.

Fades between ranges

The velocity ranges are often faded into each other. This is designed to create a smoother transition between the different velocity layers. In order to successfully fade between ranges, you need to have well-matched samples. This includes

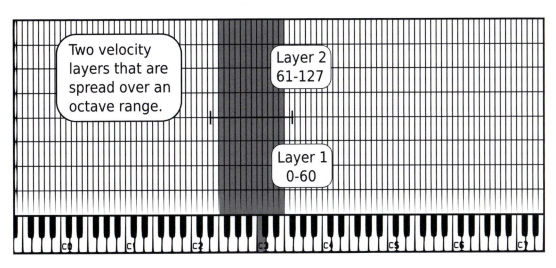

Figure 6.5 Velocity Layers.

a good match between the samples' pitches, starting points, volumes, and timbre.

Some samplers allow fades between zones that are side by side along the keyboard map as well as between the different velocity ranges.

Both fade types operate in a simple manner. You can set a point after which the sample will slowly fade out or in. Samplers that allow the fades to be set graphically are less common, but powerful. Most samplers provide knobs or sliders to do this, and you have to rely on your ear to tell you if it is correct. Samplers that have an auto fade feature set the different fades between layers (Figure 6.6) and zones (Figure 6.7), but there is no guarantee of a successful fade until you hear the results.

An example of both types of fades could be used on a sampled trumpet. Let's say the instrument is six zones wide and four zones deep. This means there are six individual samples spread across the keyboard range, and each of them has four overlapping zones with different samples representing different velocity layers (Figure 6.8).

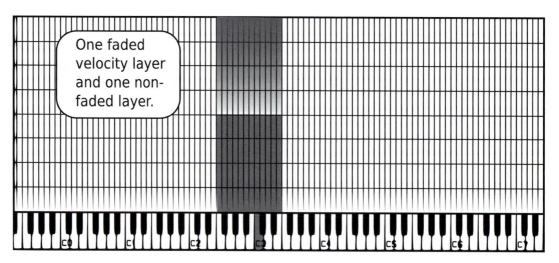

Figure 6.6 Layer Fades.

Figure 6.7 Zone Fades.

Sample #1 fades into Sample #2

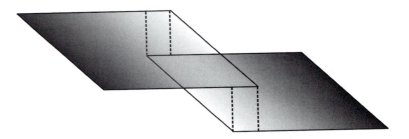

The first type of fade exists between the velocity zones. The first velocity range might be from 1 to 30 and the second range from 30 to 60. If a fade was placed between these zones, the first zone might start to fade out at 25 and be gone by 35. The second zone might start at 25 and be full strength by 35. This means that for a portion of the velocity range, both samples will be playing but never at full strength. These fades are a useful tool for creating a smooth transition between the zones.

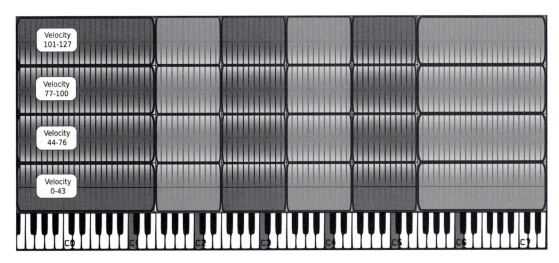

Figure 6.8 Example Instrument.

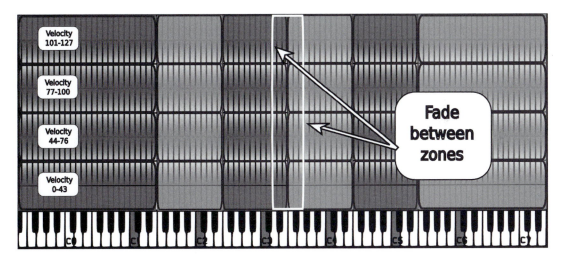

Figure 6.9 Example Fade.

The second fade type is between the zones along the range (Figure 6.9). This is a less common type of fade, but it is used in some samplers. Using the same trumpet example, these fades take place between two consecutive zones. The first might end at E3, with the next zone beginning at F3. The fade between these zones ties the two zones together and creates an overlap. When a D3 is triggered, the upper zone is also triggered but at a softer level. When an F3 is triggered, the zone below might also be triggered but at a softer level.

This is useful for creating a smooth transition within a source that has a large variance from zone to zone. If the instrument you are creating has the maximum zones, then cross-fades between them is not possible. When using a large number of velocity ranges, you usually do not need to use fades between them.

6.1.4 Using the round-robin function

Another common tool that is available when using multiple zones is the alternate or *round-robin* function. When two or more zones are layered on top of each other with matching velocity ranges, the sounds can be layered together or played in sequence with a different sample being triggered each time a note is played.

This is appropriate when striving for a natural and organic instrument. When a real source is played, there are often differences in the sound each time it is played, even when the volume is identical. Samplers that lack the round-robin function can often be identified by a listener because of the consistency from note to note. An example of this is sampled drums. If the exact same snare drum sample is repeated over and over, it sounds very repetitive and robotic. Each note triggered is exactly the same. Using an alternate sample (Figure 6.10) function, there is more variety and less repetition. Over time, the samples are repeated in a less obvious manner.

6.1.5 Looping

A universal feature that samplers offer is looping. The looping ability of most samplers is limited to several types (Figure 6.11). The loop points determine the start and the end of a

Figure 6.10 Alternate Samples.

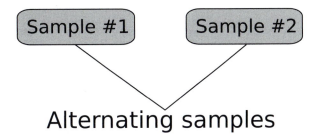

Alternating samples

Figure 6.11 Looping Options.

section of the sample that is looped. The loop parameter determines the direction in which the looping takes place.

- *Forward loop:* The sample plays from the beginning of the audio file, past the loop start until it reaches the loop end point. It then returns to the loop start point and repeats the looped section until the key is released.
- *Alternate loop:* The sample plays from the beginning of the audio file past the loop start, until it reaches the loop end point. It immediately plays backward from that point to the loop start point. Once it reaches the start point, it plays forward and backward between the points until the key is released.

Samplers are also capable of reversing the basic playback direction of the sample. This means that a sample can be played back forward (as it was recorded) or it can be reversed. Switching the playback direction of the sample does not change the basic looping functions, although switching the playback direction and looping it does give you expanded options.

The key to successful looping is to create a seamless transition between the start and end of the loop. When a sample has a change in dynamics over time, it is difficult to loop because the loop point will have an obvious difference between the beginning and the end. The reversed loop can be a solution for this, but often it seems like the sound is pulsing louder and softer. If that is the desired result, then that is okay as well. If you want a loop that seems to sustain with no obvious loop, then place the start and end points along a section of the sample that has as consistent a level as possible. Also, be aware that any change in pitch will be more obvious when looped. If there is a change in pitch, then the reversed loop might sound like a thick vibrato, but the forward loop will not usually produce acceptable results.

Another way to create seamless looping is through the use of cross-fades. At the point where the sample is looped, the audio can be set to cross-fade between the end of the loop and the beginning of the loop. This is extremely useful when you want to create smooth loops but are having a hard time finding good loop points.

When setting loop points, there are two general methods. One is to create a very short loop. This might be the length of a single cycle. The other is to use a longer section in the loop. The source will determine which of these is appropriate, but both are valid options. It is recommended that you try both shorter and longer loops (Figure 6.12) to discover which is better for your specific use. You will find that shorter loops have a very repetitive sound and longer loops are more organic and often more natural.

Figure 6.12 Long vs. Short Loops.

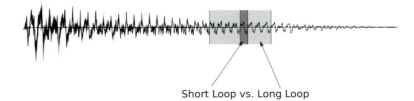

Short Loop vs. Long Loop

Figure 6.13 Good Loop Point.

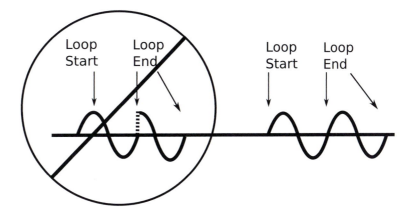

Loop Start   Loop End   Loop Start   Loop End

An important sampler feature is the ability to audition loop points and hear how they sound. A sampler might have no graphic display, so listening would be the only thing you can do in that case. If you hear an audible click at the loop point, it is most likely due to the waveform not lining up properly at a zero crossing. If the sampler has the ability to snap to the zero point along the waveform, then try this and it may fix the problem. Snapping to the zero point will ensure that the loop ends and starts at the same crossing point level (Figure 6.13).

If you have the ability to see the waveform, then zoom in and set the start and end points so when the loop repeats, the waveform follows the same trajectory. You normally have to audition a few different points until you find one that works. There are often a number of different loop settings that can produce acceptable results.

6.1.6 Envelopes

An important sampler function that needs to be adjusted during the looping phase is the volume envelope. Volume envelopes have been around longer than samplers and are used in a variety of ways in music technology. The term *envelope* refers to the volume of a sound over time and is divided into four simplified phases: attack, decay, sustain, and release (Figure 6.14). Every sound can be broken into these four phases, although most sounds certainly can be analyzed using many more divisions than four. An envelope

Figure 6.14 Possible Envelopes.

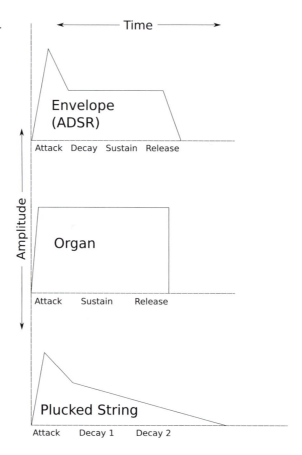

is a tool used by samplers to control the level of sounds, and you can set the attack, decay, sustain, and release (ADSR) of each sample.

The attack is the time from the start of the sound until it reaches its full level. Once that point is reached, there is a period of decay as the sound reaches the sustain level. If the sustain level is the same as the full level, there is no decay. Once the key is released, the sustain phase ends and the sound releases. For zones with no loop points, the sound plays out its length through these zones and no length is added. For zones with activated loop settings, the specific loop setting determines whether the loop will continue and fade out during the release phase or whether the loop will end and the remainder of the sample (after the end loop point) will play during the release phase.

The strength of the volume envelope model is its simplicity. Most sounds really do follow these four phases, and synthesized sound can be realistic and organic when following the same model. However, the volume envelope model is often oversimplified, in that a sound that is being sustained can often increase in volume to a point that is louder than the first attack. This is not taken into consideration in most envelope settings. Some samplers do have additional points along the ADSR envelope, and these can be used to create more complex volume envelopes (for example, A1, D1, A2, D2, S, R).

When an envelope is used to control volume, keep in mind how it affects the other parts of the sampler. A long attack time creates a long fade into the sample, and the initial attack is lost. An attack time of 0 starts the sample immediately with no fade. The decay adds a decrease in volume in addition to any natural decay in the sample. The sustain parameter affects the level of the sustained section of the sample, including looped sections. The release parameter allows the sample to finish past the loop point when the key is released.

6.2 Advanced mapping

Video tutorials

Modern samplers often have advanced mapping features that provide additional realism and creative control. These features are introduced here and there are several tutorials online that demonstrate their full power.

6.2.1 Key switching

There are a few additional features you may want to incorporate into your instrument, depending on the specific sampler you are using. Most of the features discussed so far are universal, but the following features exist in some samplers and not in others. Key switching is one of those advanced features. *Key switching* (Figure 6.15) allows a sampler to respond to a key being pressed, usually in the upper or lower range of the keyboard, which results in a switch between different sample sets. To use this function, you need a full 88-key keyboard or an additional keyboard set in the appropriate range to trigger the switch. The sample sets can represent just about any instrument, but typically characterize alternate performance styles. An orchestral bank would be a good use of key switching. If you are playing a sampled violin and you want to play staccato and legato in the same phrase without setting up multiple samplers, then a key-switched sample is a possible solution. During playback of the violin sample, you can play the key switch trigger note

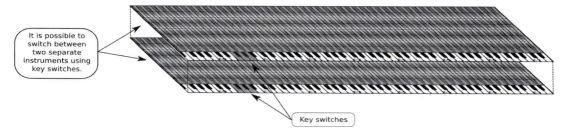

It is possible to switch between two separate instruments using key switches.

Key switches

Figure 6.15 Key Switching.

and the sampler will switch to a set of staccato sample zones. Once the legato trigger is pressed, the legato zones replace the staccato zones.

Each of the zones, in this example, could have multiple velocity layers and be fully functioning instruments. Key switching can be used in live performance and also to save time and create extremely realistic performances.

6.2.2 Scripting

Another advanced feature offered by a small number of samplers is scripting. *Scripting* consists of a computer programming–style language that gives additional control possibilities over the sampler. This ability is used in the creation of super-realistic virtual instruments. The power of this feature can be seen in performance modeling. A script can be used to filter incoming MIDI data. A keyboard is normally used to control samplers, but sometimes the instrument that a sampler is recreating is played in a different way. For example, a violin has four strings, so you can only play four notes at a time (usually only up to two are played simultaneously as a double stop). The specific notes that can be played simultaneously are determined by the tunings of the strings and the size of the performer's hands. When a keyboard controls a sampled violin, however, there is the option to play the violin the same way you play a piano: as many notes can be played simultaneously as there are fingers to play them, and there are no limitations on playing notes that are side by side. This results in a performance that can sound like a string ensemble rather than a solo violin.

A script can be programmed to translate the incoming data from the keyboard into what is physically possible on a violin. If too many notes are played at the same time, then the script can either trigger them naturally or leave some of them off completely. If programmed correctly, a script helps the performer produce realistic results. It is possible to play a sampler

157

realistically without scripting, but it takes more attention to detail on the performer's part.

Scripting can also create complex performance patches. A common example is a strumming guitar that reacts to specific keys being pressed that change the chord being strummed. The style of strumming can be changed as well. Scripting can add a number of additional features to sampling, but it has a steep learning curve. As with all programming, it is no small feat to master the capabilities offered by scripting.

Scripting example

There are several examples of sample playback software that utilize systems similar to scripting to create realistic performance functionality but that cannot be altered by the end user. Many of the available sample players in this category are created by software that allows script programming. The instruments are subsequently packaged into software that only allows playback of the samples, with no specific control over the coding.

6.3 Fine-tuning

Once the samples are loaded, the next step is to fine-tune their settings. The primary objective is to create an instrument that has the appropriate consistency from zone to zone. Within each zone, you have to adjust the loop points and loop types. You also set the root note, the zone range, the zone output settings, the sample playback direction, and velocity information. Between the zones, you need to develop consistency for each of the possible settings. If you like the velocity settings of one zone, you may have to go back and forth to the other zones to match how they sound. This is true for all of the settings: to achieve consistency, the settings must be tweaked. It isn't as simple as copying the exact settings. Even if the recorded samples sound very similar, they may have enough individual characteristics that the settings are quite different from zone to zone.

If the samples sound too different from one another, then you may need to rerecord some of them. Sometimes a good audio editor is able to change the samples until they sound more consistent, but this is often not possible. It may be difficult to match the sound when rerecording samples the second time around, due to equipment differences, temperature differences, and humidity differences. Be willing to compromise when needed.

An excellent method of checking your sampled instrument is through a MIDI test sequence. MIDI data can be sequenced to trigger every aspect of your sampled instrument. A test sequence can test each individual note though all 127 velocities. It can also test every note on a specific velocity through every range. Testing out the instrument in this way is a good method to see how consistent each zone is. Once you have run the tests, you will have an idea of what remains to be tweaked.

MIDI test sequence

6.4 An example of mapping

Video tutorials

Let's take an imaginary example through the entire mapping process. This example uses a typical set of files that might have been supplied with a music technology magazine at your local bookstore. The example comes on a CD-ROM and includes a folder that has 12 files in it. Each has the root name *Rhodes* followed by the name of the root note (*Rhodes_C3*). This is the standard naming system, and it makes it easy to map the files out. By using the file name to identify the proper root, some samplers can automatically set the root notes when the files are imported into a zone.

The first step is to import the files into separate zones. You can import one file at a time, making other adjustments before importing other files, or you can import them all and make adjustments afterward. Initially you should try importing them all at once. The sampler will either automatically set the root notes based on the file name, or you will have to set each of the root notes manually.

Once the files are imported into zones, organize them in the display so that the first sample is C1, or the lowest pitched sample. When working with a lot of samples, it is good to keep things neat and tidy. If there are alternate samples or samples to be used as additional velocity layers, it is appropriate to create groups to further organize them.

The root note is already set for each of the samples, but the zone boundaries will be set to the default, which spans the entire keyboard range. This means that when a single note is triggered, all of the samples will play simultaneously. To change this, the individual zones should be set so they do not overlap. The number of notes that each sample will cover depends on the total number of samples. In this case, the 12 are C1, F1, A1, C2, F2, A2, C3, F3, A3, C4, F4, and A4. There are three samples per octave, which translates to each sample covering about four notes.

Once the samples are arranged into zones, you can set the output routing for each zone. The default output routing is the primary stereo output, but in certain situations there are reasons to use different output settings. Surround-sound samples are the most common reason. You might also want to send the louder velocity layers out through a separate set of outputs for individualized processing. In this case, the samples are not using surround sound or multiple velocity layers so you would leave the outputs set to the default.

There are no additional samples, so there is no need to set velocity layers, round-robin settings, key switching, or any other advanced parameters. The only other features that might apply here are loop points and the envelope settings. If the sampler does not have graphic loop settings, the looping has to be done exclusively by ear. In reality, these samples might not need loops, but you may want to loop them to create an interesting effect. They definitely do not need to be looped to conserve on sampler power because there are only 12 files, and some sampled instruments can use hundreds of files without running out of processing power.

To begin looping, start with a single sample, such as C3. The sample sound has a continuous decay with no significant section of sustain. This makes it difficult to use a standard forward loop to create a natural loop, because the level has a noticeable change when the loop initiates. The alternate loop would work better, because the loop moves back and forth between the loop points and the level never jumps straight from the end to the beginning. The loop function in this case is not used to create a natural sustain but to create a slight pulse when the note is looped. The alternate loop works well for this.

Start the loop about 30 percent into the sample and hold it for nearly a third of the length of the sample. The longer alternate loop helps prevent too much of a pulse from being introduced into the sound. The setting used on this sample is the starting point for all of the other samples. You have to tweak them to make them all sound similar and to create a pulse that has a consistent tempo from zone to zone. The decay on the original recording proves to be the largest hurdle in the creation of this instrument, but only because we decided to loop the sounds. If we had decided not to loop them, it would have been a much quicker process.

In addition to the basic loop settings, we also need to adjust the overall amplitude envelope settings for each zone. The attack setting determines the fade into the sound. Use the lowest setting possible to ensure that the samples are triggered immediately. The initial decay is set to a medium setting to help the transition between the full volume and the lower sustain level. The sustain level determines how loud the sustained area of the instrument will be. Set this to a lower level than the full volume, to add a little bit of dynamic change. If you want the sound to decay slightly even after the keys are released, then set the release time to a relatively low setting, but not zero. You might also decide to change the playback mode so the sampler will play the end of the file past the loop end when the key is released. This brings an additional sense of realism instead of fading out, while still looping the same section over and over.

161

Another option when using a sampler with key-switching capabilities is to create one set of zones that has a natural decay and another set that is looped. You can switch between the two sets by triggering the specific key-switching trigger. If your sampler does not have this capability, then consider splitting the velocity range in half and putting the looped zones in one half and the non-looped zones in the other. This means that when you play louder, the samples will loop, and when you play softer, they won't. This can be tricky to accurately control in a performance setting, but it can be done. It would be easier to use in a sequencing situation, without the pressure of live performance velocity accuracy.

Taking this example further, let's add further velocity layers. With 36 samples instead of only 12, things are a little more complicated. The additional 24 samples are two more sets of the original 12 but recorded at different levels. Each sample still needs to be mapped into zones and prepared in the same way as the basic 12. The velocity information and possible cross-fades need to be set as well. Instead of treating each set of 12 individually, it helps to create three groups that can be edited as one. This can save a lot of time and help prevent any zones from being overlooked.

The velocity layer settings depend heavily on the specific samples. If the levels of the samples are split evenly, then you might be able to set the velocity layers in thirds. This is not the normal outcome, because the samples are rarely recorded that precisely. Instead, there will be some critical listening involved in order to find the proper velocity breaks. Set the velocity layers at even intervals to start with and then adjust until the results sound natural.

Once the velocity layers are close, use the test MIDI file to trigger each note of the instrument at all velocity levels in succession, and listen for even level changes through the entire velocity range. As you do this, make adjustments until everything sounds smooth. You can also set up a peak meter

Metering options explained

to measure the changes in a more specific manner, but doing this by ear is also acceptable.

If you are having difficulty creating smooth transitions between the velocity layers, you may want to experiment with cross-fades. The fades create an overlap between the velocity layers and might fix the transition problem. Cross-fades are not always the most natural solution, because it is often even harder to implement natural sounding cross-fades than to create natural-sounding velocity layer switching without fades. The simple explanation for this is that the sample files need to be near perfectly matched in sound quality and pitch to create a good fade from one to the other. Fades also work well when fading in a second sound such as a louder attack that is mixed with a softer attack. In this case, a sound is added and not replaced by a second sound. Remember that cross-fades can also be used as an effect or to create a brand new instrument by layering additional sounds into the original sample.

The instrument is now functional and is ready to be tested. During the testing phase, you should check and double-check every sample and zone for consistency. For a relatively simple instrument, such as the one described here, the process of checking everything may not be overly complex. For a larger project, this phase may take several hours or more. Even though it may take a while, it is critical to the process and will save time and energy in the long run.

6.5 Additional sampler features

There are additional features available that can be used to enhance the sampled instrument and provide extra features to the sampler. These features are used more in a sound synthesis and creation process rather than in the capture and sampling of sound sources. These additional features include envelopes, filters, LFOs, and built-in effects. These can be used both globally and with specific zones and groups.

If the instrument you are creating is designed to be a realistic copy of a real instrument, then these effects might not be useful at all. It is when the original recordings are lacking something that you might use these tools to help cover up any weaknesses that exist. One common problem that might require additional help is the lack of proper velocity layer samples. If you want the sound of a piano played very loudly but in the recording session the piano is only recorded at a medium level, then you can either rerecord the piano samples or make adjustments in the sampler (see explanation of effects in Chapter 5) to emulate a louder velocity range.

These tools can also be used creatively to make new instruments and variations on copies of real instruments. One use of a sampler can even be to emulate a synthesizer. If the same basic source waveforms that synthesizers use are imported into a sampler, it is possible to closely imitate the functionality of a synthesizer and in many ways surpass what some synthesizers are capable of doing. When it became apparent that digital sampling was practical, digital synthesizer designers realized that the same technology could aid in synthesizer design. Conventional oscillators could be eliminated and replaced with very consistent digital waveform samples. There are still true analog synthesizers available, but the majority of digital synthesizers use samples as the basic waveform source.

6.5.1 Flexible envelopes

Envelopes can control other things besides volume, such as filters and pitch. Most samplers have at least a volume envelope and an assignable envelope. The assignable envelope is able to control specific parameters that are user definable—for example, the cut-off frequency of a filter (see Section 6.5.2 for more on filters). Several advanced samplers allow multiple envelopes that can control many different sampler settings. Sometimes the envelopes are completely flexible, with a number of different user-definable points along the envelope length instead of the traditional ADSR.

6.5.2 *Filters*

Filters were described earlier in Section 5.4.2 in Chapter 5. Filters are spectrum processors that can boost or cut different frequencies. The common filter types are peak, high pass, band pass, and low pass. The typical parameter settings are the frequency knob, the resonance knob, a filter-type knob, and a keyboard-tracking switch. The frequency knob determines the shelf point of the selected filter. When used with peak and band pass filters, the resonance (or Q) knob is used to adjust the width of the filter. The keyboard-tracking switch allows the filter to react to note number or velocity information.

An envelope can be used to control the filter, primarily controlling the cut-off frequency of the filter. This can be used to slowly adjust the filter over the length of a sample. This standard feature is especially useful when creating synthesizer-type sounds and instruments.

The keyboard-tracking switch is typically used to create a more realistic response from samples that have multiple velocity layers and that change when played louder. A piano is a good example of this. When a piano is played loudly, the added force of the hammers on the strings pushes the strings further out of the natural resting state. This creates more tension in the strings. The added tension and displacement results in different harmonics and an overall shift up in the harmonic content of the resulting sound. Simply stated, the piano has a different sound when played louder, and that sound has more high-frequency content. The keyboard-tracking switch can mimic this same phenomenon by boosting high frequencies when a higher velocity is present. Of course, the same function can trigger the filter in other ways, and you can create very interesting sounds using the filter. Another example is a trumpet that is played in the upper register where it is brighter in harmonic content. Tracking the keyboard for pitch information and applying that control to the shelf-point of a filter increases the harmonic content as pitch increases.

6.5.3 LFOs

A low-frequency oscillator (LFO) produces a periodic waveform that can be used to control various parameters. The LFO itself is not typically used to create sound but acts solely as a control source. An LFO might be used to control the volume of a sampler, and the resulting sound could range from a slight tremolo to a full-range pulse. LFOs generally have a rate parameter with optional sync options, a destination selector, a waveform selector, and an amount knob.

The rate parameter sets how often the waveform will cycle. LFOs are designed with the ability to cycle very slowly, and this is used to create sweeping effects and slow pulses. LFOs can also cycle quickly and are used to create flutter effects. The rate can often be synced to the tempo of the host software that is controlling the samplers, or to an internal tempo if the sampler operates on its own. The sync allows the LFO to be timed to any other parameters that can be synced, including any available MIDI sequencers. A sync reset ability may also be included which will reset the waveform each time a key is pressed. If this function is turned off, the next triggered note may start in the middle of a waveform cycle.

The destination selector provides the possible choices that the LFO can control. These choices include volume, pitch, filter, and, in some samplers, the effects and processor settings. Most samplers have multiple LFOs, and each LFO has multiple destination possibilities.

The waveform selector provides different simple waveforms that can be used as the source. This includes a set of basic wave shapes, such as sine waves and square waves, but it might also include other waveforms such as random or specific step patterns. These can be used creatively to adjust the various destination parameters.

The amount knob sets the depth of the LFO effect when controlling a parameter. The specific limits for the amount knob varies from sampler to sampler and from parameter to parameter. When a volume is controlled through an LFO, it will range from full volume to no volume. If the amount is set to full, the volume will pulse from full volume to silence. If the amount is set to less than full, it will sound like a pulse but without the complete silence. The amount can be set very low, which sounds like a mild tremolo. The same is true for each of the destination choices (for example, pan and pitch), and the limits on the amount will vary accordingly.

6.5.4 Built-in effects and processors

Most samplers have built-in effects and processors or have direct access to them. While several of the available samplers can stand alone without additional software or hardware, there are several that are a small part of a larger system. The stand-alone samplers often have a variety of tools to add to or change the sampler's output sound. The samplers that are a part of larger systems typically do not have effects and processors built directly onboard but have access to them from the system to which they belong. These effects and processors vary from system to system, but they fit into the same categories as those listed in Chapter 5.

USING SAMPLES 7

The goal of this chapter is to give you information that will help when you are using sampled instruments. This includes a range of items, from a basic MIDI explanation to a large set of musical definitions and explanations. While this chapter has a lot of information that is not directly related to the creation of sampled instruments, it is included here because it pertains to the use of sampled instruments.

7.1 MIDI

Additional MIDI resources

While it is highly probable that you have a basic understanding of MIDI, the following overview is designed to be a good refresher while covering some key elements that apply to sampling techniques.

7.1.1 Overview

MIDI (musical instrument digital interface) is a data communications protocol invented to allow different musical devices to communicate with one another. No audio is transmitted via MIDI, just messages that instruct audio devices what to play. There are many types of data that are carried via MIDI, such as "note on" and "note off" messages. Other data types that are transmitted are control change, continuous controllers, and system exclusive data.

The initial specification for MIDI was released in 1983 to allow synthesizers to transmit data back and forth with one another, as well as with personal computers.

7.1.2 Connections

MIDI cables, unlike most computer peripherals, carry data in only one direction. This is different from the bidirectionality found in serial, USB, and FireWire cables. There are usually three types of MIDI ports on a MIDI device: MIDI In, MIDI Out, and MIDI Thru. MIDI In is the port that receives MIDI note on, off, and velocity messages, as well as all other incoming MIDI data. MIDI Out transmits the MIDI data from the controller to either the computer or another MIDI instrument. MIDI Thru is a special MIDI output that replicates the MIDI data received at the device input. This allows several MIDI instruments to be daisy-chained together with one main source. To daisy-chain two or more devices, you simply connect the MIDI Thru on the first device to the MIDI In on the second device. You can do this with as many devices as desired, but there is a delay involved which becomes an issue when three or more devices are connected. The MIDI standard allows for a total of 16 different channels to be carried via the MIDI cable. This means that there can be 16 separate parts with note on and note off messages, set to play on different instruments. The cabling for MIDI connections entails plugging the MIDI Out or Thru of one device into the MIDI In of the desired target device. A MIDI cable is connected from the MIDI Out of one device to the MIDI In of another device. The MIDI data flows from MIDI Out to MIDI In.

7.1.3 Interfaces

MIDI interfaces are devices that can be connected to your computer via either a serial or USB connection. These devices then provide the MIDI ports that can be connected to your various MIDI devices, such as controllers or sound modules. Drivers may need to be installed on your computer to recognize and communicate with the various MIDI interfaces. Read the manufacturer's manual to see if a driver must be installed. Traditionally the MIDI keyboard controller is then connected to the MIDI interface, which can be routed to your sequencing program and then back through the MIDI interface to your chosen musical device. With today's computer technology, MIDI data can be routed to internal software instruments as well.

Modern MIDI keyboard controllers can connect directly to a computer with a USB connection. No additional MIDI interface is needed. These controllers are both the MIDI interface and controller in one single unit. When your computer recognizes these USB controllers, it is seeing them as a MIDI interface. Quite often these controllers have additional MIDI Ins and Outs to connect to other external MIDI devices.

Every computer operating system (OS) has its own way of dealing with the routing of MIDI devices, so users need to make sure that they configure their devices appropriately according to their sequencing software recommendations, as well as the recommendations of the OS developer.

7.1.4 MIDI data

MIDI data messages are transmitted serially through a MIDI cable, meaning that 1 bit follows the previous bit. Each message is in an 8-bit word, or byte. With each word, however, the first bit indicates whether it is a "status byte" with a 1, or a "data byte" with a 0. This gives MIDI data a 7-bit resolution, so the true range of MIDI data is a total of 128 points of resolution, 0 through 127. What this means is there are 128 playable notes and 127 velocities, with 0 equivalent to a "note off" command.

Pressing a single key of a MIDI keyboard controller sends three 8-bit messages: note on/channel number, note number, and note on velocity. Releasing the key sends a similar set of three messages: note off/channel number, note number, and note off velocity. In most situations, a note on velocity of 0 is the equivalent of a note off message.

Because of the density of MIDI messages that are sent and received, it is possible for a note message to be "stuck" on. This is fixed by either pressing the key again to retransmit the note off message, or by pressing a "panic" button featured on most controllers and sequencers. This sends a note off message to every note on every channel of the system and takes a few seconds to complete.

7.1.5 *Controllers*

The term *controller* is used in different contexts with regard to MIDI. The controller can be the entire MIDI keyboard. The various faders on knobs on a keyboard can also be referred to as controllers. It is important to note the context in which the term is being used.

MIDI data can also carry control change information. The control knobs and sliders on a MIDI keyboard transmit this control change information. This is one context in which these input devices are referred to as controllers. These controllers are used to adjust a particular parameter of an instrument. They are merely a different means of inputting MIDI data. Just as there are a total of 128 MIDI note numbers, there are also 128 basic MIDI controllers. Some of these controllers are predefined if the device is designated to receive such information. For example, controller #1 is the modulation wheel, which is found on most standard MIDI devices. The modulation wheel sends a 0–127 message when the wheel is moved up or down. It can then be designated in the target MIDI device to adjust a specified parameter, such as filter cutoff frequency or volume.

It is important to note that the majority of control change data is sent only when the controller is moved. It does not continuously transmit data. For example, if the controller is set in the middle, when the user loads a patch, whatever parameter is being adjusted by the controller is set to its default position according to the patch. However, if the controller is moved slightly up or down, the parameter immediately jumps to the data that the controller is transmitting. This can create an abrupt jump in whatever parameter is being controlled.

MIDI keyboard controllers, which have various sliders and knobs, have a default number assignment, which transmits on that previously assigned controller number assignment. This controller number is often adjustable by navigating menus on the controller itself or via editing software on the more advanced MIDI controllers.

In the same way that controller numbers are adjustable on the keyboard controllers themselves, the parameters that the user wishes to control can also be adjusted to react to a specified controller number. With many software instruments, they can be attached to the controller number by specifying that they be attached to the next controller number message that they receive. Once the controller is moved, the parameter is "learned," and so this feature is often called *MIDI Learn.* The controller controls the associated parameter until further changes are made.

Since MIDI sequencers record MIDI data, such as note on, velocity, and note off data, they can also record controller data. It is generally helpful to record the controller data on a different track of the MIDI sequencer, so that the controller's data can be muted or edited easily. Sequencers allow the user to quantize and edit the MIDI data, and controller messages can be edited in the same way. Most sequencers allow controller data to be drawn in with various tools, or even automated as a virtual fader on the sequencer itself.

MIDI has many more capabilities than the basic operations mentioned here, including transmitting time code for synchronization and system exclusive messages for storing patches with an editor/librarian. The MIDI standard changes occasionally to incorporate new features as recommended by the MIDI manufacturers, but the basics of transmitting note and controller information have remained the same. It is an open system with room to grow, which is why the format has survived for decades. It can be as simple or as complicated as the user requires.

7.2 Musical instrumentation

Employing sound sampling to assemble the parts or voices of a performance ensemble is an extension of the same challenge faced by musicians for centuries. The musical success of a composition is largely dependent on the choice of voices or instrumentation.

Here are definitions of the typical choices for instrumentation.

7.2.1 Emulation

Emulation refers to the performance of sampled sounds as exact replacements for traditional instruments. The objective is to convince listeners that they are hearing an acoustic performance of the same music.

7.2.2 Transcription

Transcription means setting an existing piece of music for a different ensemble of instruments. The original form is followed quite closely. The new ensemble can be of acoustic, electronic, or sampled instruments, or any combination. The performance characteristics of the new ensemble should be faithful to the composer's original intention.

The music stays the same; the ensemble is different.

Listen to Vladimir Ashkenazy's 1982 orchestral transcription of Mussorgsky's "Pictures at an Exhibition." Compare it with the original piano version performed by Vladimir Ashkenazy.

7.2.3 Realization

Realization means setting an existing piece of music for new electronic or sampled/processed instruments. It is a form of transcription usually describing electronic orchestration.

The music stays the same; the ensemble is of electronic instruments.

Listen to Isao Tomita's 1975 synthesized electronic realization of Mussorgsky's "Pictures at an Exhibition." Compare it with the original piano version performed by Vladimir Ashkenazy.

7.2.4 Arrangement

An *arrangement* is a form of transcription that allows for greater flexibility in modifying the form and instrumentation of the original composition.

The music is recognizably the same; the ensemble can be the same or different.

Listen to Emerson, Lake, and Palmer's 1971 arrangement of Mussorgsky's "Pictures at an Exhibition." Compare it with the original piano version performed by Vladimir Ashkenazy.

7.2.5 Composition

Composition means the act of creating a new musical work for acoustic, electronic, or sampled instruments. A thorough knowledge of the performance characteristics of each voice of the performance ensemble is an integral part of effective composition.

7.3 Composition and orchestration

Eventually you must create a plan and a prescription for the composition to achieve a maximum of ingenuity with a minimum of resources.

7.3.1 The plan

Composition is the organization of sound. The composer can start with a specific scene, musical form, event timeline, or other organizational plan for which a musical/sound design establishes the points of emphasis. The challenge is to select suitable musical relationships that support the demands of the plan and to reject those that are not appropriate.

The choice to avoid one of the traditional forms can result in a *soundscape* or sonic environment (as opposed to "music"). The challenge here is to control elements of the environment that will attract attention that the composer does not want.

Music and sound design share similarities in the way they evoke a human response to the organization of sound. (The composer Edgard Varese used the term *organized sound* to describe a viewpoint of his electronic music.) It is helpful to look at these similarities in order to anticipate and predict the natural human response to organized sound. These tendencies are discussed later in the section on observations (Section 7.8).

7.3.2 The prescription

Orchestration is the selection of sonic elements for participation in the performance of musical thought. There is a unifying strength to establishing a fixed ensemble of sounds and restricting any unsupportive sounds that do not belong to that ensemble. There are participants in the sonic ensemble that fulfill the traditional roles of soprano, alto, tenor, and bass performers. There is a predictable clarity to the voicing of an ensemble where the soprano and alto voices can move faster and can be stacked closer together than the tenor and bass voices. If you want the sound texture to get thick and muddy, then stack lower voices close together.

The process of musical expression has several steps. Sampling is a way of assembling a performance vehicle, an ensemble, a sonic palette. It is common and natural that in the process of sampling and the manipulation of samples, a melodic, harmonic, or rhythmic musical idea is discovered. These discoveries can be the seed of the organization, development, or expansion of the sonic resources. Manipulation of sampled sounds is one way of getting to know the full resources of the ensemble. Sound sampling, of itself, is not a finished composition. It is only a step in identifying the ensemble, exploring the features of an ensemble, and then participating in the performance of a musical idea.

You don't really know the resources available inside a sound sample until you take it through all of the manipulations. Play it forward/backward, faster/slower, expanded/contracted, normal/inverted. The manual scrubbing of tape across the playback head can provide an interesting irregular speed aspect worth investigating. The alteration of direction of a short motific element can reveal some interesting rhythmic and harmonic outcomes. The popularity of "scratching" a phonograph record, as practiced by current DJ artists, is a renewal of this historic practice and has revealed the worthy intrigue of this manipulation.

Let the sound samples inform you and inform the composition through exploration and development. At times it is good to drift and wander through the resources—you might discover an intriguing element generated by a manipulated sample. The challenge is in deciding what to select from the inventory of sound samples that serve the composition best and what to leave out. Eventually you must create a plan and a prescription for the composition.

7.4 Musical terms

The artist can choose to use sampling to construct a single instrument, an ensemble of instruments, or a sonic environment or soundscape. The use of sampling to construct an ensemble of sounds or to broaden the sonic palette for a musical composition or sound design does not change the basic interactions that occur in organized sound. The music community has found labels and descriptions for sonic elements that have understandable functions in a composition.

The following is proposed as a lexicon of common musical terms that can be applied to sound designs and non-tonal music. These terms can apply as well to abstract sonic environments or tonal music. They are discussed here to reinforce the observation that listeners relate to organized sound in predictable ways. It is helpful for this discussion to identify these characteristics as being *horizontal* if we are considering their progress through time and *vertical* if we are stopping time to evaluate relationships at a single point in time.

7.4.1 Melody

Melody is a horizontal element. It changes in pitch (frequency) over a period of time. Typically it describes a single pitch row. The human mind has a unique ability to remember melodic designs and melodies for long periods of time. Many cultures rely on melodic recognition and melodic memory for songs and chants important to religious or cultural history. The shortest recognizable pitch row of a melody is called the *motif*. The motif can be an organizing element in a composition, and development of the motif can give listeners identifiable references to the composer's choice of form. In the sonata allegro form, the melodic intrigue is found in the presentation of a motif, the composer's exploration and development of this material, and then a return to familiar ground: statement, departure, return or exposition, development, and recapitulation. In a fugue, the melodic intrigue is found in developing the motif in different voices, as a conversation. There were rules for the construction of a fugue that became a popular intellectual exercise in the classical music period.

7.4.2 Harmony

Harmony is a vertical element. It denotes the association of multiple pitches at a point in time. The musical terms *consonant* and *dissonant* are useful for describing harmonies that are pleasing or at rest, or harmonies that are displeasing or in motion. Patterns of dissonance and consonance in harmony can create tension/relaxation, motion/rest—in other words, they can also create rhythm. The harmonic progression is important in identifying a phrase.

7.4.3 Rhythm

Rhythm is a horizontal element, signifying the accents, beats, or pulses occurring over time from patterns of tension and relaxation. Changes in amplitude, pitch, or texture can contribute to tension or relaxation, motion or rest.

7.4.4 Texture

Texture is a vertical element. It describes the choice of instrumentation that results in tone color. The individual waveform or complex waveform resulting from combinations of voices contributes to a sonic texture. When viewed from the horizontal perspective, changes in texture can create tension and relaxation, and thus rhythm. Additional observations about texture are offered in Section 7.8.

7.4.5 Motif (or motive)

Motif (or *motive*) is a horizontal element and designates the shortest recognizable pitch row of a melody. Even soundscapes can have an element that functions as a motif. This short, recognizable statement can be one of the primary organizing features of a composition. It is the statement of this motif, and its evolution through the composition, that provides melodic intrigue.

7.4.6 Phrase

A *phrase* is a horizontal element. It refers to a complete musical statement, or to the region of melodic or harmonic progression that appears to come to rest or relaxation. It is similar to a sentence in grammatical construction.

7.4.7 Form

Form, a horizontal element, is a collection of phrases that results in a complete musical thought. It is similar to a paragraph in grammatical construction. In musical and literary construction, there is logic to forms that utilize "statement—departure—return" or "exposition—development—recapitulation" as an organizing template. This is the musical form of *sonata allegro*, which is widely used in pop songs and in classical music. There are other musical forms, such as the round and the fugue.

7.5 Compositional techniques

The development of motific (thematic) material is the real craft of the gifted composer. Following the manipulation of thematic materials becomes the intriguing journey for the listener.

Here are some of the traditional manipulations that composers utilize to develop melodic materials.

7.5.1 Rhythmic expansion

Rhythmic expansion means increasing the duration of a musical element—for example, changing each quarter note to a half note. In signal processing, this is comparable to slowing down the playback media. Slowing down the playback mechanics lengthens the playback time. A byproduct of this analog manipulation is a lowering of pitch in direct proportion to the change of speed. In digital manipulation, these two elements can be separated. A digital change of playback speed need not be linked to a digital change of pitch. In MIDI performance, controlling the data is not associated with the sound sample, and therefore need not affect the pitch.

7.5.2 Rhythmic contraction

Rhythmic contraction means decreasing the duration of a musical element—for example, changing each quarter note to an eighth note. In signal processing, this is comparable to speeding up the playback media. Speeding up the playback mechanics shortens the playback time. A byproduct of this analog manipulation is a rising of pitch in direct proportion to speed. Digital manipulation separates the elements of frequency and time into separate components; they are not mutually interactive.

7.5.3 Intervallic expansion

Intervallic expansion means increasing the distances between melodic steps. This is a very difficult manipulation to do in analog processing. However, with a computer it is quite easy to perform the necessary mathematical calculations and to impose numerical manipulations on numerical data. Transposing the data that represents a pitch row only requires the computer to multiply each numerical representation by a linear number. Changing the intervallic steps of a pitch row only requires the computer to multiply each numerical representation by an exponential number.

7.5.4 Intervallic contraction

Intervallic contraction means decreasing the distances between melodic steps. It is the opposite of intervallic expansion and functions similarly but in the opposite way.

7.5.5 Inversion

Inversion denotes changing the direction of intervallic steps. If the interval went up, inversion moves it down; if it went down, inversion moves it up. This is like a mirror reflection of a melody based on intervallic distance from a reference center pitch. This can be organized by diatonic interval, linear distance, or logarithmic distance changes. Many MIDI software editors allow you to specify or interpolate this manipulation.

7.5.6 Retrograde

Retrograde signifies the performance of the melodic materials in reverse direction, starting at the last note in the row and proceeding to the first. This is like a mirror reflection of a melody based on note performance order. In analog signal processing, this is comparable to playing a tape backward. In sound sampling (recording), the onset of the sound is at the release portion of the envelope, and it progresses through the sustain, decay, and attack in reverse order. In MIDI performance control, there is no change in the envelope of the sound because the MIDI data does not include the sound; rather, it contains the performance data.

7.5.7 Sequence

A *sequence* is a statement of a theme or motif that is presented in the same voice but starting at a different pitch. This is different from the use of the term *sequencing* when it is applied to "event data sequencing." A *sequencer* is an event recorder/player.

7.5.8 Imitation

An *imitation* is a statement of a theme or motif that is similar to the original. It can be presented in a different voice, at a different interval, in a modified (but recognizable) form, or it can be built of the same shapes.

7.5.9 Repetition

A *repetition* is a statement of a theme or motif that is identical to the original. The use of echo and looping creates repetitions of material. Be careful that the use of repetition does not make the development predictable to the point of being boring. If you want a sonic line to move into the background and become unnoticeable, then use exact repetitions of the line. It will not be long before the listener becomes familiar (even bored) with this line and moves it to the background of their attention.

7.6 Performance characteristics

In the human performance of musical instruments, there are traditional nuances that add warmth, emphasis, individuality, or urgency to the expression of a line.

Here are some typical performance nuances.

7.6.1 Vibrato

Vibrato is a periodic change in frequency (pitch). The human performance of musical instruments or of the voice is usually characterized by a smooth vibrato, sinusoidal in wave shape. The use of LFOs in signal processing offers the opportunity to select any available wave shape for vibrato.

7.6.2 Tremolo

Tremolo is a periodic change in amplitude or timbre (harmonics). The human performance of musical instruments or of the voice is usually characterized by a smooth tremolo, sinusoidal in wave shape. The use of LFOs in signal processing offers the opportunity to select any available wave shape for tremolo.

7.6.3 Articulation

Articulation describes the amplitude profile, or envelope—that is, the changes in the amplitude and harmonic content of a sound as the performer exercises deliberate performance techniques such as bowing (stringed instrument), tonguing (a wind or brass instrument), striking (a percussion instrument), or plucking (a string instrument). The simplest amplitude history of a performed sound can be defined by the nature of the attack, decay, sustain, and release components of the envelope.

7.6.4 Dynamics

Dynamics denotes the amplitude profile of a phrase or passage. Fixed dynamic levels establish the balance or mix of the ensemble to ensure that the melodic and supportive elements are in the correct relationship. Gradual changes in amplitude (crescendo or decrescendo) add performance nuance and tension/relaxation to the expression of a line.

7.7 MIDI performance control

The digital control of musical instruments provides great flexibility, accuracy, and variety to the performance of sound samples. When performance control is via MIDI, the samples can be stored in either analog devices or digital devices.

As MIDI is a system for recording, storing, editing, and recalling performance attributes, it must be considered separately from the sound sources. Its control features apply to sound sources of analog or digital synthesis, and to sound effects as well as musical sound samples. The source of sound is not the domain of MIDI control; it is the execution and modulation of those sounds.

The musical keyboard is the most common MIDI controller, but there are many other controllers. The keyboard is the most visual of controllers, so it is used here as the model for other controllers.

The following are some of the MIDI messages that represent performance attributes:

7.7.1 Note on/off

An event's onset is triggered by a switch closure (depressing a key). The end of an event, *note off*, is marked by the release of the key. The key must be released and depressed again to extract the next note on event. This is a problem when emulating the bowed or plucked string (violin or guitar, for example), where a new note onset can be achieved by reversing the direction of the bow or pick stroke. The number of physical movements required of the performer is changed by a factor of 2.

7.7.2 Note number

The *note number* message indicates the assignment of the chromatic range of the keyboard to specific note numbers. Transposition of sampled sounds is easily achieved by recording the designated sample at one note number and then executing its playback by a different note number. This is particularly advantageous in performing sound effects, as a transposition alters the perceived size of the effect in direct proportion to the transposition. The sample of a gunshot from a pistol can sound like a cannon when transposed down. Likewise, the chirp of a sparrow can sound like a lion.

7.7.3 Velocity

The *velocity* control extracts the numerical value that represents the amount of time elapsed during the keystroke from the top to the bottom of its excursion. The assumption is that the faster a key is depressed, the louder the desired performance. The MIDI velocity value is typically used by applying this control value to the amplitude domain.

7.7.4 After-touch

After-touch is activated by a pressure sensor at the bottom of the keystroke that produces a variable control value. The harder the performer presses on this sensor, the larger the numerical value. When applied to the frequency domain, it produces a pitch bend. When applied as a gate for a low-frequency control waveform, it introduces a vibrato. When applied to the timbral (harmonic) domain, it changes the initial filter shelf point.

7.7.5 Breath control

Breath control is initiated by a pressure sensor held between the teeth that allows the performer to articulate a note much as a brass or woodwind player would tongue the same note. This MIDI data can be applied to a filter for harmonic content changes or to an amplifier for volume changes.

7.7.6 Patch change

Patch change is a MIDI message that contains the instruction to change the selected voice to another voice. The term *patch* comes from the early days of analog synthesis, when patch chords were necessary to establish the construct of a voice of a synthesizer. A map or facsimile of this patch chord routing could identify a structure for a selected violin patch, flute patch, and so on.

7.7.7 DSP

The *digital signal processing power (DSP)* of the modern computer can be used to modify the performance attributes of analog/digital synthesis, wavetable synthesis, or sound samples. The developed skill in analytical listening will serve you well in creating sensitive musical performances from MIDI controllers.

7.8 Observations

The following observations are made with regard to appropriate and realistic sample usage. Incorporating these thoughts into your compositional process helps the samples come alive and creates realistic results.

7.8.1 Observations about timbre

An ensemble of voices built of the same or similar waveforms functions as a homogeneous choir. A single contrasting voice built of a distinctively different waveform can be heard clearly against this homogeneous choir. This is why a single oboe in a symphonic orchestra can cut its way through an ensemble of violins.

From a complex ensemble of sounds, human perception tends to identify the highest and lowest sounds in the ensemble most clearly. It is therefore important that you take care in constructing the top and bottom voices of a texture. Interior voices of the ensemble seem to blend together unless they are exaggerated by contrary motion, contrasting motion, or by unique waveform.

7.8.2 Observations about melodic development

The first time we hear a melodic fragment (motif), we listen and remember. The second time we hear the same fragment, we compare it with the first and observe whether it is the same, similar, or different. If it is the same, we label it as a *repetition*. The third time we hear the same fragment, we compare it with the first and second and observe whether it is still the same, similar, or different from the first two. If it is the same, we begin to lose interest and possibly predict that a fourth and all subsequent appearances will also be identical. We move it into the background and look for other intrigue in the music. If it is similar or different (yet recognizably related to the original motif), we are intrigued by the development it has undergone—it has grown, matured, or evolved. We are inclined to look forward to the continued journey through the development of the melodic materials.

7.8.3 Observations about melodic motion

Providing clear identification of individual voices adds overall clarity to a composition. The identification of a voice in a polyphonic texture can be clarified by providing contrasting motion for that voice. A voice in motion can be identified against a background of voices of limited motion or at rest. A voice at rest (or limited motion) can be identified against a background in motion. A voice moving up in pitch can be identified against a background that is moving down (and visa versa). This is called *contrary motion*.

Anticipation, or the early or delayed arrival at a predicted melodic gesture, can add an element of surprise and intrigue. The listener is inclined to predict the arrival at a similar phrase length or cadence because of the composer's use of melodic sequence. Even soundscapes can have phrase lengths and repetitions that can become predictable, so consider the anticipation or delay of an arrival point as a way to add intrigue. Loops are certainly predictable. Consider altering the length of the third or subsequent repetition if you want to call attention to a line.

7.8.4 Observations about layering sounds

Modern recording and sequencing technologies provide the opportunity to layer audio tracks with sampled sounds or to create entirely new sonic instruments. Historically orchestrators of acoustic and electronic ensembles have understood the advantages of doubling (layering) voices for appropriate and desirable effect. There is a different result when the choice is made to double a voice in unison or at several octaves apart. Sometimes the objective is to create a combination voice that is different from either of the two individual voices. Other times the desire is to select particular points of accent to exaggerate the envelope of the primary voice.

Consider the limitations of the envelope and waveform of the primary voice(s) and select a complement in secondary voices that will exaggerate or enhance the desired voice. You can alter the resultant attack, sustain, or other articulation by layering a voice that has the missing components. You can call attention to a line by creative layering. Consider avoiding this when your objective is to hide or blend a line into the background.

Here are some traditional examples of doubling (layering) sounds to achieve a desired result:

1. Doubling the first violin voice with a mallet instrument (a marimba or xylophone, for example) adds percussive attack to the envelope and high-frequency content to the harmonics.
2. Doubling the string choir with an organ (no vibrato or tremolo) contributes a fixed intonation center, smooths out the differences in vibrato, and regulates the sustain portion of the envelope.
3. Doubling the string bass voice with piano or bass trombone adds percussive attack to the envelope and broadens the harmonic content.
4. Punctuating the bass line with a bass drum provides an exaggerated attack to selected points in the line. It is neither generally necessary nor appropriate to double the entire line, but only the points that need additional accent. Doubling the entire line would dilute the benefit of selective punctuation.

Sound effects for film and video are rarely constructed from single samples of a gunshot, a door closure, or a car crash. The artistic construction of several layers of sounds appropriate to the event results in far more believable effects. The objective of *audio sweetening* is to exaggerate the sound effect and make it larger than life. Listen to the sound effect of a human fist hitting a person's jaw as exaggerated in film effects and compare it with the actual sound (don't try this at home; you'll hurt someone). The sound effect artist (or Foley artist) will usually layer something like the "swish" of a rod through air, and a baseball bat striking a suspended side of beef, with the cracking of a whip for this exaggerated sound effect.

7.8.5 Observations about the string ensemble

One of the most commonly sampled choirs is the string ensemble. The traditional string ensemble is recognized as a warm and pleasing sound and is called for where large symphonic textures are appropriate. There are several elements to the original ensemble that give it its unique texture. The prescription of re-creating a traditional string choir is best served if the following observations are taken into account.

Each individual string player provides a similar waveform, yet has a slightly different intonation, vibrato rate, depth, and cycle. In the aggregation of multiple string players into a string section, the modest differences in intonation and vibrato result in the chorusing timbre desirable in the string pad. If all players in the ensemble play with perfect intonation, with the same vibrato rate and depth, the ensemble loses its unique texture.

The physical placement of the string ensemble provides separation of the first violin, second violin, viola, celli, and bass sections (left to right). This placement assists in localization and clarity of the traditional ensemble. The choice to constrict this placement into one or two locations (left and right stereo) reduces the clarity of each section's voice leading.

Articulation of the bowed string is important to the authentic re-creation of stringed instruments. In a live performance, the players use many different bowings to express the line authentically. Most sampled reconstructions are limited to one articulation, thus dampening the believability of the performance. If keyboard velocity and after-touch modulations are available, the application of velocity to amplitude changes and of after-touch to timbral modulations will approximate some of the additional expressive articulations normal to string performance. As a bowed string is played louder, it not only gains in amplitude, but it has an altered amplitude profile of the harmonic content in the waveform.

It would be advantageous to have a separate sample for each of the different bowings. Using MIDI control to engage layered samples of the various harmonic profiles can also add realism to the expressive performance. Breath controllers provide this benefit to the performance of wind and brass samples.

Consider multitrack recording of each individual section in the string ensemble if the objective is a traditional ensemble. This allows you to phrase and articulate each line separately. Also, the shift from four or more lines performed from a sample-playing keyboard to one line is not an accurate representation of multiple lines converging into ensemble unison of the same number of players. If the sample is of 8 violins, then the 4-voice polyphony would represent a string ensemble of 32 players. The single note would be of only 8 players, not the true 32 players in unison. The mix down affords the opportunity to adjust each section's appropriate dynamic expression and physical localization. Even in a nontraditional performance, the clarity and musicality in performance of each line benefits from this individual attention and performance.

7.9 Bibliography

Adler, Samuel (2002). *The Study of Orchestration.* 3rd ed. W.W. Norton, New York.

Alten, Stanley (2007). *Audio in Media.* Thomson Wadsworth, Belmont, CA.

Aldoshina, A., and R. Pritts (2005). *Musical Acoustics and Engineering Applications.* Compositor Pub., St. Petersburg.

Eargle, John (2005). *Handbook of Recording Engineering.* 4th ed. Springer Verlag, New York.

Meyer, Jürgen (2004). *Akustik und Musikalische Aufführungspraxis.* PPV Medien GmbH; Auflage: 5, aktualis. A. (Juni 2004).

Ottman, Robert W. (2000). *Elementary Harmony: Theory and Workbook.* Prentice Hall, Upper Saddle River, NJ.

Rogers, Bernard (1970). *The Art of Orchestration, Principles of Tone Color in Modern Scoring.* Greenwood Press, Westport, CN.

Piston, Walter (1947). *Counterpoint.* 1st ed. W. W. Norton, New York.

EXAMINING THE
ROOTS OF SAMPLING   8

Timeline

The desire to re-create the performance of existing instruments from a keyboard has an interesting history. Early pipe organs would use Fourier additive synthesis in combining multiple pipes and resonators of various sizes and designs to emulate brass and woodwind voices. These were not samplers, but the intrigue of orchestrating at the keyboard was obvious.

Keyboard performance of recorded samples was advanced in music production by Harry Chamberlin in the 1950s with his keyboard instrument that engaged the playback of a bank of tape samples of the chromatic range of selected instruments (string ensemble, choir, and the like). The mechanics of this instrument were complex in that each tape sample contained the attack, decay, and sustain history of the performance, and it had to be played from the beginning in order to be an accurate representation. At the onset of a key, the tape was drawn past an individual tape playback head (one for each note) and continued for the length of the tape sample. At the conclusion of the length of the tape, the sound would stop. At the release of the key, it would rewind to the start and await the next key depression. The Mellotron followed the Chamberlin and was a licensed product of the Chamberlin patents.

An interesting continuation of this technology was the "L.A. Laugh Machine" which was a proprietary product used to create audience responses to television shows where applause, varieties of laughter, and other audience sounds were prepared as samples for keyboard execution. It is still common practice to sweeten audience responses with crowd samples in television production, live sporting events, and concerts.

The human interest in exploring new sonic resources, excluding sound synthesis, began with the first technologies that provided recording, manipulation, and performance of sound. The wire recorder, phonograph, optical film, and tape recorder became exciting tools for sonic exploration. The recorded sounds could be manipulated by altering the playback mechanics to change direction, speed, regeneration, and other properties.

There are several historic labels that have been associated with these early explorations. They typically took place in broadcast or recording facilities because of the high cost and limited availability of quality audio recording and playback equipment. Modifications to the mechanics of such equipment would allow for nontraditional playback of recorded samples and loops. Great skill and creativity was necessary in tape splicing to assure artistic musical performance.

The organization of such manipulated sounds into credible compositions still remained the challenge of the composer. The job was not finished just because there were clever sounds generated; now the composer faced the job of organization.

8.1 Timetable of processes
8.1.1 1950s–1960s
Classic studio technique

The *classic studio technique* typically recorded samples from existing musical instruments. The Columbia-Princeton laboratory in New York was a recognized center for this exploration. Manipulation by tape splicing, playback reversal, tape speed manipulation, and tape looping were common practices for developing the expanded sonic inventory. Composers such as Vladimir Ussachevsky and Otto Luening were notable. At Ottawa University in Canada, Hugh le Caine and Norman McLaren were pioneers. Hugh le Caine's "Dripsody"

(1955) was a composition created from a sample of a single drop of water and is an excellent example of manipulation by the classic studio technique.

Musique concrete

In *Musique concrete*, sound sources were typically recordings sampled from natural environments. The broadcast facility at Radiodiffusion Television Française (RTF) in Paris was a recognized center for this exploration. Composers such as Pierre Schaeffer and Pierre Henry were notable.

Electronic music (Elektronische Musik)

Sound sources for *electronic music* (or *elektronische Musik*) were typically recordings sampled from electronic sources. Electronic oscillators, microphone feedback, and ring modulation were employed to generate basic sonic materials. Manipulation by electronic filters, echo/reverberation/regeneration, and classic studio technique were employed to develop the composition. The radio facility at Norwestdeutcher Rundfunk (NWDR) at Cologne and Bonn, Germany, were recognized centers for this exploration. Composers such as Werner Meyer-Eppler and Karlheinz Stockhausen were notable.

The radio facility of Radio Audizioni Italiane (RAI) at Milan, Italy, was also a recognized center. Composers such as Luciano Berio and Bruno Maderna were notable.

Early electronic music composers found that manipulation of recorded sound materials opened a whole new palette of sound sources for musical expression. These were sound samples in the purest sense. They gathered recordings of selected sounds, stored them in available recording media, and then manipulated these samples upon retrieval for musical performance.

By using mechanical manipulations of recorded sound, the traditional compositional techniques for developing musical materials could be applied to nontraditional sounds.

8.1.2 1970s–1980s

Digital samples

The ability to record audio information into digital files on a computer changed many of the opportunities for sound sampling and performance. The properties of digital recording, editing, and reconstruction expanded the tools for composition and orchestration.

Early computing systems had limited bit depth, sampling rate, and random access memory. The quality of an audio recording is proportional to these elements. The lower the bit depth and sampling rate, the less accurate the resultant waveform. The length of a sample was directly proportional to the amount of memory that could be assigned to the sample.

A change of the pitch of a sample was achieved by increasing or decreasing the playback sample rate. This result was the same as changing the playback rate of mechanical (tape) storage in that the pitch, amplitude, and harmonic change history of the sample changed in direct proportion to the speed. It was quite easy to recognize the resulting distortions to the accurate sound of a sample that was altered by more than a major third. The chromatic range of an instrument was typically divided into zones or splits, where five or eight zones of the range were assigned to separate samples of the original instrument and then transposed up and down to fill the zone. There was a noticeable seam where the zones joined and where the sample of one zone that had been sped up met a zone where the next note had been slowed down. The number of splits or zones was dictated by the amount of available memory. The greater the number of zones, the less exaggerated the transposition of neighboring samples. Ray Kurzweil offered the best solution to this early problem by using *floating point* sampling and compansion of the digital sampling process. The Kurzweil 250 and K1000 samples were noted for their realistic piano, choir, and string ensembles because they utilized as many as 50 samples of the chromatic range of these sounds. Each sample needed to be transposed only once (a half-step up or down).

The amplitude history (envelope) of a sample was difficult to reproduce by early computing systems. The length of the envelope was directly proportional to the amount of memory. Sampling rate and bit depth both needed memory that subtracted from the amount of time that the sample could be recorded. Musical instruments having relatively short envelopes, such as percussion, were not as difficult to reproduce as were instruments with a long envelope such as piano and violin. The exceptions in the percussion family were the cymbals, which had long envelopes and very complex harmonic histories. Roger Linn produced several popular drum sample players in the early 1980s that provided very realistic performances of a studio drum kit.

For samples that required a sustain portion to the envelope, looping of an interior portion of the envelope provided an option that conserved computer memory. Just as in tape splicing, the loop point (or splice) had to be of the closest possible amplitude and harmonic content to be successful. A poor choice of a loop's splice point resulted in a click or pop at each repetition.

If an instrument did not normally have a sustain feature in its performance (piano, for instance), it was possible to artificially create sustain by looping. The problem here was that by artificially increasing the ever-decreasing amplitude of the piano envelope to make an "equal power" loop, the background noise floor was also increased. (This demonstrates why it is so important to take great care in controlling the noise floor and ambience while recording samples that may be processed by looping.) Later computers and more elegant software provided "auto splice" features that located and adjusted the ideal silent splice points.

8.1.3 1990s

Wavetable synthesis

Progress in digital wave construction made it possible to record and analyze the harmonic history of a sample and then construct a digital waveform with the same history. A table of these samples can be stored as the digital representation of an instrument or sample. Musical instruments have a very complex harmonic and amplitude history during the performance of a single note. There are significant changes taking place throughout the note's performance. The success of a *wavetable* is based on the accuracy of the documentation of these changes.

1. A single wavecycle uses only one cycle of a sampled sound. A table of the relative amplitudes of the harmonic content of the sample sound is stored, and then computer software is used for digital reconstruction, repetition, and transposition. Software synthesizers generally use this process.
2. Multiple wavecycles use more than one cycle of the sampled sound, allowing for a more realistic performance history of the early timbral changes. Sounds are held in read-only memory (ROM) of a sound card. Looping is still necessary to construct a performance note of reasonable length.

The weakest part of present-day sampling technology is the accuracy of the harmonic profile of a performance note.

8.2 Timetable of historically innovative systems

8.2.1 *Analog devices*

Date	Device	Creator	Characteristics
1946	Chamberlin	Harry Chamberlin, Upland, CA	Precursor to Mellotron Recoil tape reels for each key Provided attack, initial decay, sustain Sustain time limited to length of tape At release of key, tape would rewind; rewind time proportional to length of key depression
1963	Mellotron MK1 L.A. Laugh Machine (Black Box)	Licensed from Harry Chamberlin	Proprietary sample player providing sound effects, applause, laughter, and audience response samples for TV and film sweetening

8.2.2 *Digital devices*
1970s–2000s

Date	Device	Creator	Characteristics
1978	Fairlight CMI	Fairlight's Peter Vogel, Kim Ryrie	Large Computer Music Instrument
1981	E-Mu Emulator	E-mu's Scott Wedge, Dave Rossum	First dedicated sampler
1982	LinnDrum	Roger Linn	First commercially successful drum machine; digital samples
1983	DX-1	Decillionix	Five-voice sampler for Apple II computer
1983	K250	Kurzweil	2 MB of ROM sample playback
1985	AKAI S612	AKAI's David Cockerell	12-bit sampling, 6-note polyphony W/MD280 DRIVE for quick-disc storage First affordable rack mount sampler Limited memory
1985	Prophet 2000	Sequential Circuits	12-bit sampling, 8-voice polyphony Much larger 512K memory
1986	ESQ-1 player and Mirage sample recorder/player keyboard	Ensoniq	8-bit sampling
1986	S900 sampler/player	AKAI	Rack mount, 12-bit sampling, 8-note polyphony
1987	MT-32 and S-10 and S-50	Roland	Combined sampling with synthesis processing
1988	M1	KORG	Sampling in a workstation
1988	S1000 stereo rack mount	AKAI	16-bit sampling
1990	K1000	Kurzweil	Sample player keyboard
1990	Wavestation	KORG	Sampling, synthesis processing, wavecycle, and wavetable applications.
1991	General MIDI (GM)	Consortium of manufacturers	Standard established for assignment of drum sounds to specific MIDI note numbers
1998	Sonic Foundry ACID	SoundForge, now Sony	Loop-based sampling and sequencing software for Windows platform
1999	Triton	KORG	64-note polyphony, workstation
1999	Reaktor 2.0	Native Instruments	Sampling for PC or MAC
1999	ESI 2000	E-mu	64 voice sampler with 4 MB sampling (expandable to 128 MB)
2003	Fantom-S	Roland	Sampling workstation, smart media storage, 16 performance pads

Note: See this text's website for a list and feature comparison of all current samplers.

8.3 Bibliography

Manning, Peter (2004). *Electronic and Computer Music*. Oxford University Press, New York.

Russ, Martin (2004). *Sound Synthesis and Sampling*. 2nd ed. Focal Press, Burlington, MA.

Miranda, Eduarto Reck (2002). *Computer Sound Design: Synthesis Techniques and Programming*. 2nd ed. Focal Press, Woburn, MA.

THOUGHTS ON THE FUTURE OF SAMPLING   9

Audio sampling is a technique that is used in nearly every area of music. From composers to pop artists, sampling is used as a means to save both time and resources. Sampling has also come into its own as a creative tool that is used to create new sounds and form brand new instruments. From the early tape-based samplers to the advanced modern samplers, a foundation has been laid for future expansion and development.

When sampling became economically feasible and widely used in recording studios, many people feared that musicians would be replaced. I'm sure the idea has crossed many producers' minds. Over time, samplers have changed the way that many things are done. In some styles of music, they have replaced musicians (Figure 9.1). In other cases, samplers have expanded what is possible and helped musicians realize their artistic vision (Figure 9.2).

Instead of producing an extreme change, samplers have become integrated into the workflow of the music creation process. The primary question is not whether sampling has a place as a creative tool (it has proven that it does), but what will happen next.

One of the most consistent trends in digital technology is the constant rise in processing power coupled with the decrease in physical size. The same holds true for samplers. MIDI production studios of the past (yes, they actually had their own studios apart from recording studios) required large amounts of space and a lot of planning and cables (Figure 9.3).

Figure 9.1 No Musicians?

Figure 9.2 Studio Musicians.

Due to the size limitations and other factors, MIDI studios and recording studios stayed relatively separate for many years. But once software samplers and synthesizers were developed and streamlined, things began to change. Computer-based digital audio systems came of age with extremely powerful systems, and the worlds of audio record-

Figure 9.3 MIDI Studio.

ing and MIDI production fully collided. One of the most popular recording platforms of the 20th century, ProTools, managed to avoid developing advanced MIDI functionality on par with many other companies. Starting in 2003, ProTools expanded on its basic MIDI capabilities (available since 1991) and finally introduced advanced MIDI features already present in the majority of sequencers. In 2007, ProTools released a new sampler called Structure. This comes more than 50 years after the original Chamberlin tape sampler. The release of a new sampler, when it seemed so late in the game, speaks volumes about the continuing trend and popularity of sampling technology.

Are there sampling avenues that have not been explored? The basic sampler design has not changed significantly in a very long time. There have been performance boosts and developments that allow more voices and larger numbers of individual samples. More advanced feature development, such as scripting, hasn't spread to every sampler. The near future will bring more of the same feature optimization. Samplers will become easier to use. Graphic interfaces are already becoming sleeker and more functional, and they will continue to do so. There will be more voices available, and the sampled instruments will become more detailed. Samplers will, however, stay fundamentally the same.

9.1 Modeling and convolution meet sampling

Modeling is the new wave of the future. Modeling is not a form of sampling but accomplishes similar end results. When creating a model of an instrument, many measurements are taken. These measurements are used to create a software instrument that simulates a sound source using mathematical modeling. Modeled instruments can reproduce these real instruments very accurately. In the future, it will be possible to create indistinguishable replicas. One difference between sampling and modeling is that, with modeling, the end user's creativity is limited by the sounds that are included with the modeled instrument. There are parameters that can be tweaked by the end user, but the fundamental sounds and programming cannot be altered. The amount of software development and programming that goes into a modeled instrument's creation is enormous. This precludes most musicians from creating their own modeled instrument from scratch.

A part of modeling is the convolution process. *Convolution*, in relation to sampling, is the ability to transfer the characteristics of one sonic item to another. It is possible to capture the essence of a concert hall's reverb and use it in a recording studio on a mix (this process is explained in more detail in Section 5.4). Convolution techniques are a part of the future of sampling. The best thing about convolution is its accessibility: almost anyone can create convolution samples. Not only is it relatively simple to sample the reverb of a given space, but it is not much more complex to capture the characteristics of other pieces of equipment.

An example of this is sending a sweeping sine wave through the filters of an analog synthesizer such as the classic Mini Moog synthesizer (Figure 9.4). Once the resulting wave is

Figure 9.4 Mini Moog.

List of software packages

recorded, the original and resulting recordings can be used to create an impulse response that describes the sound of the filter. This process is similar to sampling. You are taking a sample of a piece of equipment's characteristic and providing a way for it to be used without having the original present.

Traditional sampling, modeling, and convolution techniques are already beginning to be used together. There are software packages that give access to all three. It is very possible that these are going to grow even closer together. Imagine the potential sound possibilities that could result from such a marriage of these available technologies.

9.2 Future uses of sampling

Sampled instruments are currently being used nearly everywhere that music production is happening. The chances of hearing a sampled instrument in a given day are almost 100 percent. From cell phone ring tones, to movie soundtracks and the radio, samples are everywhere. I'm not sure this is going to change anytime soon. In fact, it is widespread enough currently that it is hard to imagine that overall usage could actually increase.

Figure 9.5 Recording Studio.

If there is one realm of the music industry where usage of sampling might continue to increase, it's in the area of home and portable studios. Amateur musicians and recording engineers alike now have access to great-sounding samples that can be used to create demos and records. In the past, high-quality instruments were expensive and out of reach for most people. Sampling technology has brought a lot of musical possibilities into the hands of the general population. The elite recording studios of the world no longer have the monopoly on great sound sources (Figure 9.5).

This doesn't mean that everything coming out of home studios is going to start sounding good. It just means that the possibility is there, if a person is willing to learn how to turn the available sounds into something great.

9.3 Summary

Samples are going to get better and better. They are going to be easier and easier to use. New technology is going to bring forth more powerful and realistic-sounding sampled instruments. All the available instruments in the world will someday fit in the palm of your hand, and they will sound just like the real thing. It is beginning to happen already, and the trend is going to continue for years to come. In this process, it is most important that creative control be left in the hands of the end user. Musicians should be able to create their own instruments. The future is bright for sampling.

INDEX

Oct. 20/08